6/86

A Theory of Technology

A Theory of Technology

CONTINUITY AND CHANGE IN HUMAN DEVELOPMENT

Thomas R. DeGregori

THE IOWA STATE UNIVERSITY PRESS / *Ames*

TO MY MENTORS *David Hamilton*
C. E. Ayres
Wendell Gordon
Melvin Kranzberg

Excerpt from *Humankind Emerging,* 3rd edition, by Bernard G. Campbell, Copyright © 1982 by Little, Brown and Co. Reprinted by permission.

The text in this book was printed by Iowa State University Press, Ames, Iowa, from camera-ready copy provided by the author.

Library of Congress Cataloging in Publication Data

DeGregori, Thomas R.
    A theory of technology.

    Bibliography: p.
    Includes index.
    1. Technology—Social aspects. 2. Technology—
Philosophy. I. Title.
T14.5.D46 1985   303.4′83   85–11800
ISBN 0-8138-1778-1

CONTENTS

ACKNOWLEDGMENTS

Everyone from whom I sought assistance in preparing this book responded in ways that were positive and helpful. Whether the request was small or large, all provided support to the best of their abilities. To recognize all who helped would be beyond any reasonable space limitations. At minimum, a few need to be singled out for special note.

The Agency for International Development (AID) has been particularly helpful in giving me the opportunity to acquire practical field experience in technology transfer and development. I have also had the good fortune to participate in seminars and other interactions with longtime practitioners in development and in science and technology's role in this process. First among those with whom I have worked is Jim O'Connor, to whom the phrase "generous to a fault" is truly applicable. Hank Miles is a friend who has always shared his ideas with me and given me the opportunity to develop mine. Randall Thompson and Rufus Long have been co-workers in the field and sharers of long discussions on technology and the specific needs of project development. Jim Ray, of the Bureau of Census and a fellow worker on an AID project evaluation, was very supportive of my musings on the theory of technology.

The United States Information Agency (USIA) was instrumental in my presentation of the ideas in this book to business groups, government officials, academics, and development planners on four continents. Interaction with all of these different professionals, peoples, and cultures was absolutely essential for gaining diverse perspectives on the technology issues I am exploring. The intellectual interchange with the many foreign service officers and other agency employees has been as valuable as the seminars and meetings in which I have been involved. Though there are far too many people to thank in this regard, I would like to say that Hugo Bayona, the program officer for science and technology, did an out-of-the-ordinary job in finding the right places for my programs. Without his efforts, I would not have obtained the intellectual benefits of participating in USIA programs.

In Pakistan, where I have had the privilege and pleasure

to lecture, participate in conferences, and engage in the development of technology, there are many to whom I am indebted. In particular, M. M. Qurashi and M. A. Aslam have been regular sources for ideas, information, and critical thinking. Two of my colleagues at Denver Research Institute, Jim Frasché and Steve Force, worked closely with me in Pakistan (and in Denver) and helped clarify my ideas on the application of theory to practical problems of technology.

Obtaining all the relevant research materials is not always easy, even when one has access to first-rate libraries. Even an author trying to keep faith with the reader by having the research current when the final copy is submitted will have difficulty gaining access to recently published materials. The staff of the University of Houston library was particularly helpful in ordering and monitoring books so that I received them almost as soon as they arrived in the library. The reference desk computer and the personnel operating it were essential in my locating books just published and in checking on bibliographic information. Interlibrary Loan filled in the cracks on materials not available in our library. I would like to thank Helen Tatman, Bill Jackson, Judy Myers, and Darrell Parkin. Also helpful in obtaining the latest publications were editors of scholarly journals, popular magazines, and newspapers for which I review. In particular I thank George Christian, the book review editor of the Houston Chronicle, for the many odd and sometimes fascinating books that I have read and reviewed for him.

A number of scholars and friends from around the country have read and critiqued this manuscript. Particular mention is necessary for Charles Louis, Anne Sutherland, Odin Toness, Loren Neill, Tom Gray, Joseph Bousquet, Jerry Wagstaff, and Tom Haskell. Patricia Pando made a substantial contribution to the initial formulation of a basic theory of technology and to the writing on the similarities in the relationship between technology and tools, and between language, signs, and symbols.

On my own campus, the University of Houston, many people answered requests for bibliographic sources on a particular topic; several read the manuscript and made important corrections and improvements. One colleague, Ken Brown in Anthropology, not only recommended and lent sources but also read the manuscript and suggested improvements. Moreover, he shared his ideas on the distinction between the origins of farming and the domestication of plants. Though he has not yet published all of these ideas, he generously allowed me to use them, which I did in Chapter 1. Brown's colleagues in Anthropology, Russ Reed, Norris Lang, and Dolph Widmer, helped me as well.

From other departments came assistance for which this tribute is barely adequate. Help came from Steve Mintz, Loyd Swenson, Sally Vaughn, and Tom O'Brien in History; William Simon in Sociology; John Coffman in Geography; Elbert King in Geology; Douglas Dykes in Chemistry; Jack Matson in Engineering; and Vince Hegarty in Human Development and Consumer Sciences. All of my colleagues in Economics were helpful in one way or another, including covering my classes when I was off giving lectures on technology. Of particular help was Richard Bean, who answered odd questions even when I called at odd hours.

My wife, Gayle, converted my atrocious handwriting into typescript and showed infinite patience, which is sometimes required of spouses during a writing or similar project. The patience of my children, Alice, James, and Roger, is also appreciated.

A special thanks is due to the combined efforts of a human being and a machine. Linda Cox took all the typescript and occasional handwriting and entered it into a word processor. With good humor she took my continual tinkering and changes and recorded them. With many demands on her time, she always seemed to turn out my work with extraordinary accuracy. In a work on technology, it seems appropriate to be grateful to machines. The computer at the library reference desk, for instance, was just one of several library computers essential for the creation of this work. Chapter 6 cites the work of others on the development of printing and its contribution to accuracy, scientific advancement, and individuality. Those scribblers who remember the days BWP (before word processor) can only marvel at this magnificent machine. As one of the world's worst proofreaders of his own material, I appreciate a machine that allows me to correct errors and enter changes throughout the manuscript and have a clean copy in the time it takes for the printer to put it out.

Last, but not least, my thanks goes to the staff of Iowa State University Press, from the acquisitions editor to the production and promotion people. Judith Gildner and Kathy Glenn-Lewin edited the manuscript, changed the style of footnoting, and overall, made the book far more readable and understandable without the least sacrifice in content.

It is traditional at this point to absolve all those who helped of any responsibility for any remaining problems with the book. This I happily do.

INTRODUCTION

Chapter 3
final chapters

For two centuries or more, a belief in the idea of progress has been shared by peoples in Western cultures. Science and technology held center stage as the prime cause of progress. For many in the United States, the institutionalization of progress through scientific and technological advance was a defining characteristic of the American way of life. Inventiveness was a trait in which Americans assumed themselves to be dominant. In the United States and in Western culture generally, however, there is an equally long tradition of dissent against the equating of technology with progress and of both with human betterment. These doubts about the essential beneficence of technological change reached a crescendo in the late 1960s and 1970s. The attacks against the prevailing views of technological progress became detailed and fundamental. They concerned the use of technology not only in the industrial countries but also throughout the entire globe.

This book directly accepts the critics' challenge. I argue that the impact of technology on human life is a primary and continuing cause of human betterment. There is no attempt made to revive any eighteenth century (or later) theory of progress. Rather, the attempt is to understand more clearly the nature of technological processes, and it is in terms of such understanding that this defense of technology is fashioned. The merits and demerits of technology have long been debated without a clear conception of what technology is. The assumption has been that we all know what we mean when we use the word technology. A major purpose of this book is to generate a clear, coherent, consistent definition of technology and then to apply this conceptualization to the issues of technology and society.

The incompatibility of modern technology with life processes has been a constant theme of critics of technology. In this author's judgment, not only is this attack not warranted by the facts, but in addition, a study of evolutionary processes provides vital insight for a theory of technology. To this end, Chapter 1 takes a layman's look at some of the theories of the origin and development of life on earth. From this inquiry are derived some generalizations that are considered equally valid toward a general theory of technological evolution. This inquiry continues with an overview of the interaction between tools, technology, and human evolution. The study of tools and human evolution allows for application of the generalizations derived earlier, and for the further development of an evolutionary theory of technology.

Chapter 2 takes a brief look at the impact of technological change on selected episodes in history. The attempt is to obtain a clearer understanding of the historical process of technology transfer. In so doing, a distinction is made between the universal aspects of technology, since it is based upon the physical principles of the universe, and the particularity of technology, because it is a problem-solving process in which its applicability is relative to the culture and the environment.

Chapter 3 sets forth 30 basic principles of technology and technology transfer. These principles form the core of a generalized theory of technology and were formulated with an eye to being operational for the process of selecting technologies for development. As principles or theories of technology they do not tell us what technologies to use, but they do tell the kinds of questions we should ask in order to select the best technologies and obtain the maximum benefit from them.

Much of the criticism of modern technology has been bound up with apocalyptic visions of doom and destruction. Fortunately, many of these visions specified the time and terrible consequences that awaited us. Chapter 4 investigates the earlier forecasts of catastrophes and finds them woefully wanting in terms of predictive powers. The 1970s was the decade that the apocalypse did not happen, though we were told with certainty that it would.

Part of the continuing concern about decline and doom involves the exhaustion of natural resources. Chapter 5 marshals evidence that natural resources are becoming more abundant. This is counter to what common sense would indicate. Further, the argument is made that we are unlikely to ever exhaust resources. The empirical argument offered is one that is generally accepted by economists. The reason that the evidence is overwhelmingly against the commonsense position is that natural resources are not "natural"; they are created by science and technology. This is one of several places demonstrating that a theory of technology facilitates the understanding of a problem or issue.

Many of the romantic critics of the Industrial Revolution establish an opposition between science and technology on one side and the humanities and the arts on the other. Chapter 6 shows the ways that technological change contributes to the growth and development of the arts. Similarly, technology transfer can be, and historically has been, a source of diversity and creative potential for cultures. The destructive potential of very rapid technological change or transfer is recognized, yet it is maintained that economic and technological stagnation is not an effective way of preserving traditional cultures; cultures are the product of past change, including technological change. The best way to preserve traditional themes and styles is to utilize new technical means to synthesize the exotic and indigenous themes and give new expression to the unique qualities of a people.

Agriculture is an obvious area in which science and technology have been beneficial. Ironically, it is an area in which critics of technology have been most vociferous. Chapters 7 through 10 explore the issues of agriculture, population, food supplies, and nutrition. The theoretical framework used for resources--namely that they are created by technology--here is applied to agricultural land. Data show that we are not running out of agricultural resources and that food supply has been growing faster than population. Most indices of development for the economically advanced nations, as well as for the third world, have reflected favorable changes of a speed and magnitude unprecedented in human history. While in no way denying that substantive problems of malnutrition and famine remain, the attempt is made to show some of the potential that science and technology offer for continued agricultural development.

The final chapters are concerned with the issues of values, the choice of technology, and technology's impact upon the quality of life in modern society. The focus is on the dynamics of technology; the argument is that the choice of some technologies facilitates the ability, through development, for continuing to choose.

In this book, from the first chapter onward, the thesis is that life in general and human life in particular have evolved not by living within limits but by overcoming them. The alternative-technology policy prescription of living within limits is ultimately self-defeating. By overcoming limits, we live longer, better lives. The sustainable society sought by many is the one that gives vent to the creative powers of human beings, to fashion the resources and overall means of life, through advances in technology, science, and the arts. The purpose of a theory of technology is to enable greater under-standing of the way technology developed and of the continued direction of technological change. The goal is to obtain the maximum benefit from it.

Antitechnological theories and policies can be destructive for the human enterprise. So can mindless protechnology advocacy. The understanding of the technological process as a resource-creating human activity is an essential condition for sustaining and enhancing human endeavor.

PART I

Technology and Evolutionary Principles of Change

CHAPTER 1
Technology and the Process of Life

Now that technology policy has become a central issue of debate among economists and other practitioners of development, the scope of discourse has broadened enormously. Technology policy is no longer concerned almost exclusively with expanding economic output. Issues concerning the selection of technology and its impact upon women; minorities; local, regional, and national identities; and the distribution of income have become as important as economic efficiency and increase in production. Concern has been raised about adverse cultural change, deleterious environmental impact, and irrevocable biological destruction. The discussion of technologies has led to discourses on future lifestyles, new eras or ages of humankind, and the exploration of cosmic issues such as the nature of life and physical processes. There are calls for more spartan lifestyles; and the second law of thermodynamics is used to argue that affluence must pass as energy and raw materials for production inevitably and ineluctably decline (Rifkin 1980). Emphasis is placed on renewability, consistency with nature, and living within environmental limits. Smallness has been argued to be a biological virtue (Sale 1980). Clearly it is better to be discussing the noneconomic implications of technological change than to treat technology as mere gadgetry. Further, there are criteria for assessment in addition to economic efficiency and growth in output. However, many of these large issues of biology and physics (entropy) appear to be questionable as to immediate development decisions and technology choices. The challenge to development theory and practice is serious and sufficiently widespread that it cannot be ignored or summarily dismissed. The following brief exposition on the origins and development of life is meant only to indicate that certain concepts about the use and development of technology are not inconsistent with life as we know it on this planet. That many of the critical episodes in the early development of life bear remarkable, if not superficial, resemblance to contemporary problems may provide useful analogies for understanding, but does not close out inquiry concerning the nature of our contemporary difficulties.

3

ORIGIN AND DYNAMICS OF THE LIFE PROCESS
     Earth was formed about 4.5 billion to 5 billion years ago,
and life is believed to have begun on it between 3.2 to 3.8
billion years ago (Levin 1978, 249; Ponnamperuma 1981, 125;
Washburn 1981, 147). The life process can be defined as striving
for survival by extracting energy from the environment and using
it to make self-preservation more possible. The totality of the
life process strives toward continued reproduction. Any other
goal is self-defeating (Seeds 1981   Day 1984, 257-61). Another
author notes that "life is also a chemical system that interacts
with its surroundings; after life originated, it had a rapid,
widespread, and irreversible effect on the atmosphere, the
surface water, the soils, and the sediments" (Skinner 1981, 1).
The chemistry of life as we know it is based upon the almost
unique bonding properties of carbon and its ability to form with
other elements complex organic molecules (Rosa 1982, 36-37; Seeds
1981, 346).
     Fundamentally, life can be understood as a system with four
basic constituents. They are (1) proteins, (2) nucleic acid, (3)
organic phosphorus compounds, and (4) cell membrances. "Proteins
are essentially strings of comparatively simple organic molecules
called amino acids." These are defined more fully in chapter 9.
The nucleic acids include DNA and RNA. The organic phosphorus
compounds "serve to transform light or chemical fuel into the
energy required for cell activities." The cell membrane
"provides a relatively isolated chemical system within the cell
and keeps the various components in close proximity so that they
may interact" (Levin 1978, 250).
     The very nature of life itself necessarily creates stuff
that we have come to call waste and/or pollution. "Pollution is
not, as we are told, a product of moral turpitude. It is an
inevitable consequence of life at work" (Lovelock 1979, 27,
108-9). Life takes energy from the environment and releases it
in a degraded form, i.e., the process of entropy. Though the
energy that passes through life confirms the second law of
thermodynamics (entropy) by passing out in a less usable form,
life itself (as many have noted) is an island of negative entropy
in that it builds complexity. The very chemical processes of
life involve the taking in of compounds from the environment,
breaking them down, using part, and expelling others. These
parts expelled can be defined as waste. Under other circumstances the wastes of life can be, and in the past have been, a
creative environmental factor for the further development of
other life.
     The first life forms developed in an atmosphere that is
thought to have been a hydrogen and methane environment with a
small amount of oxygen in it (Lovelock 1979, 18; Cloud 1981, 7;
Washburn 1981, 148). The organogenic elements that covered Earth
were methane, ammonia, carbon monoxide, water, carbon dioxide,
hydrogen sulfide, and hydrogen. Without an ozone layer to
filter it, ultraviolet light was bombarding these molecules,
converting them into amino acids, sugars, nucleic bases, and

other carboxylic acids (Calvin 1981, 21). A number of famous experiments conducted in this century have shown that organic compounds could have been formed from the action of ultraviolet light upon the early atmosphere of earth (Oparin, Haldane et al. experiments in Cambridge Encyclopedia of Earth Sciences, 349-3527). The next stage in the process is life, and there are a number of competing theories on how it came to be on Earth.

These early life forms are part of the large category of procaryote. They were unicellular and were reproduced by cell division. They were heterotrophic in that they could not manufacture their food and were forced to live off already existing organic materials. These organic compounds were converted into energy by chemical processes of fermentation. These life forms were anaerobic, meaning that not only could they reproduce without the presence of free oxygen; an oxidizing environment would have broken down the chemical composition of life as it then existed.

A system of life consisting entirely of heterotrophs is inherently limited and theoretically doomed to extinction. No matter how large the preexisting stock of organic compounds (i.e., food) was, the procaryotic life forms eventually would have exhausted them. Some new organic compounds were probably being created, and living cells were undoubtedly consuming other living cells, as they have been doing ever since. Further, any oxygen buildup in the upper atmosphere in the form of ozone would have begun to shield out the ultraviolet rays that created the organic soup that fed life as it then existed. Thus, the ozone layer that was later to become the protector of life was initially an agent for the destruction of the food supply for the then existing life forms.

The first food crisis was resolved by the evolutionary processes that led to the creation of photoautotrophs in the form of photosynthetic procaryotes, possibly as early as 3 billion years ago (Levin 1978, 252). A photoautotroph is a life form that can use the sun's radiant energy (in this case the most abundant wavelengths) to convert carbon dioxide and water into glucose (as a storehouse of life's energy) and oxygen. This process is written in most basic works on the subject as

$$6 \ CO_2 + 6 \ H_2O \xrightarrow{\text{sunlight}} C_6H_{12}O_6 + 6 \ O_2$$

When organisms take in oxygen to oxidize their food (glucose or the $C_6H_{12}O_6$ in the above equation), they reverse the process and acquire energy for living. It is chlorophyll in the organism that absorbs light and traps its energy in a form that can be used (Carola 1981, 128-32).

We humans frequently find that when we solve one problem we at the same time create another. Progress in the human endeavor is most meaningfully defined not in terms of the ultimate or final solution to problems but in creating smaller or less important ones than those we solve. In life, the development of photosynthesis among procaryotes meant the generation of more of

that deadly, life-threatening substance, free oxygen. Initially,
there were large amounts of oxygen acceptors such as ferrous iron
that could serve as an oxygen sink, keeping atmospheric oxygen
down to tolerable levels (Levin 1978, 252; Cloud 1981, 11;
Chapman and Schopf 1983, 302-20). Eventually, the most readily
accessible oxygen acceptors acquired a full complement of oxygen,
and oxygen began to build up in the atmosphere. Meanwhile,
procaryotic life was retreating to environments protected from
death-dealing oxygen. The reducing environment became an
oxidizing one. This process may have taken as long as a billion
years (Cloud 1981, 11).

About 1.3 billion to 1.6 billion years ago, a new life form
appeared, the eucaryotes (Cloud 1981, 15). The first eucaryotes
were photoautotrophs like some of the procaryotes. Their most
important characteristic in the context of this analysis was that
they were aerobic organisms. The ozone layer that prevented the
creation of organic soup now shielded and protected the new life
form. The free oxygen in the environment that was destructive to
anaerobic procaryotes became a vital component of the new life
processes.

If prevailing theories are correct, life on earth overcame
the food crisis followed by a toxic oxygen crisis through
evolutionary changes. There is nothing in these theories that
argues that the evolutionary changes had to occur. They did. If
they hadn't, we wouldn't be here to write about them. It is
not beyond the realm of possibility that elsewhere in the
cosmos (once or many times), similar biological challenges were
faced with extinction occurring in some instances and diverse
successful options pursued in other instances.

The emergence of photosynthesis and eucaryotes had evolutionary advantages beyond that of survival. Photosynthesis and
the creation of oxygen allowed for a more efficient converter of
energy by oxidative metabolism in eucaryotes (Gest and Schopf
1983, 135-48). To William Day (1984, 257, 263), what the
eucaryotes achieved by respiration was a release of available
energy from glucose that was 18 times greater than their
predecessors were able to obtain. The procaryotes "were on the
edge of a larger dimension and created an energy potential that
could propel a greater form of life into it." For procaryotes,
it was "an energy potential they could not handle." For Peter
Decker (1981, 529), greater energy efficiency is a defining
characteristic of evolution. "Since life thermodynamically is an
open system, in all theories on its origin, the energy source
gains first importance. Other things being equal . . . evolution
. . . primarily should be proportional to the sum of energy
turnover per mass unit."

Anaerobic respiration yields about 7% of the energy that
would be possible through oxidation of glucose. The ethanol
given off in anaerobic fermentation also contains energy that,
for the organism, is waste. These processes are used today in
digesters to create methanol for fuel from animal manure and
vegetation.

This greater energy efficiency created surplus energy

potential and opened up new evolutionary possibilities. Sexual differentiation and mitotic cellular division opened possibilities for genetic variation and the acceleration of evolutionary change (Washburn 1978, 151; Cloud 1981, 137; Levin 1981, 252-53; Skinner 1981, 1). It is often stated that "the price of sex is death." Only chance mutation changes simple cells that otherwise reproduce by dividing themselves and perpetuating exact copies. For more complex forms death is necessary to create the possibility for new life and the evolution of new life forms (Mallove 1984). To survive (in some cases to the present), many procaryotes established symbiotic relationships with (or within) eucaryotes. According to Margulis (1981, 33), "the aerobic metabolism and genetic systems of these 'higher cells' led to the eventual dominance of the most complex eucaryotes, the metazoans, and the green plants." Dominance is not necessarily numerical. To the extent that it may involve a value judgment, it is implicitly and explicitly accepted in this book. About 600 million years ago, eucaryotes with "hard coverings and skeletons in a number of animal groups" evolved and with them the fossil record proliferates (Skinner 1981, 1). After about 3 billion years of evolution, life as most of us nonbiologists recognize it took hold in the form of multicellular plants and animals.

THE LIFE PROCESS AND DEVELOPMENTAL POSSIBILITIES

There are a number of lessons for economic development, or at least insights into the process, that we can derive from the study of the development of life:

1. Life as we now know it created the conditions for its own existence with the assistance of early life forms. Humans have complemented the life-sustaining conditions through the development of technology that has made environments habitable for their kind. In most human habitats, the conditions for even bare existence, let alone the level of existence for high population density, are dependent upon human modification of the environment.

2. What appears to be fixed, absolute limits to development can be overcome. There were food supply and pollution crises in the past. Life overcame them and established a new framework for growth and change. Similarly, in the human endeavor, this book will show how science and technology has overcome and can continue to overcome barriers to development. For the life process, we cannot assert that these limits necessarily had to be overcome, merely that they were. It took time--in some instances, eons--and possibly chance. For humans, technology has possibilities for future problem solving that are neither certain nor automatic. It requires the application of effort and intelligence. The only point here (which shall be reiterated later) is that apparent limits to growth should not be used as a device to close out inquiry or foreordain technology and development strategies.

3. Entropy, or the second law of thermodynamics, applies only to closed systems. Thus, life as a whole is definably an island of negative entropy. As long as there is available energy outside the life system, then there is the continued possibility to move in the direction of greater complexity and diversification. Obviously, all technologies, large scale or small, are subject to the laws of thermodynamics. If current theories are correct, entropy as applied to the known universe will mean playing out the available energy sources in the far distant future. Entropy in this larger sense cannot be used as an argument for certain kinds of technologies over others. As Lovelock has correctly and beautifully stated, "The death sentence of the Second Law applies only to identities, to closed systems, and could be rephrased: 'Mortality is the price of identity.' . . . In the end, the sun will overheat and all life on earth will cease, but that may not happen before several more eons have passed. Compared with the lifetime of our species, let alone that of an individual human being, this time span is no tragic brief spell, but offers almost an infinity of opportunities to terrestrial life (Lovelock 1979, 125).

4. The life process itself produces waste. Again, as Lovelock states, "The first appearance of oxygen in the air heralded an almost fatal catastrophe for life [but] ingenuity triumphed and the danger was overcome . . . by adapting to change and converting a murderous intruder into a powerful friend" (p. 31). Science and technology are the means by which humans turn adversity to opportunity. Again, we would not want the statement on the possibility of solving today's environmental problems through technological change to give rise to complacency or the polyannaish belief that technology will automatically solve our problems. While the "blind" forces of evolution may have solved the problems of eons ago, today our time frame is shorter and concerted organized intelligence must rise to the tasks at hand.

5. While the life process on earth may have created the conditions for its own existence, "nature" as popularly used (presumably meaning the ecosystem apart from humans) is not a "friend" to any particular species or group of organisms. Again, from Lovelock we learn that "by far the most poisonous substances known are natural products," with some of the most deadly being "entirely organic products and but for their toxicity would be suitable candidates for the shelves of the health food store." Nature can mean climatic variability--drought, floods, disease, and other natural disasters. Nature is! When we humans modify it to suit our purposes, as we do by merely existing, we must do so with care and with foresight, based upon our scientific knowledge.

6. There appears to be a similarity, probably superficial, between biological evolution and technological evolution. Certainly one cannot rigidly apply the rules derived from the study of one of the processes to the other. However, the similarity does raise further questions as to the thesis of incompatibility between technology and continued human life. Both processes move from simple to complex, though some simple

forms continue to exist.   Both processes seem to accelerate
through time.   Roughly 2 billion years, or over half of life's
existence, was limited to simple procaryotes.   Eucaryotes have
existed for less than half, about 45% of life's existence.   Over
65% of this time was prior to the creation of hard surface or
skeletal creatures.   This latter group, then, has only existed
about 15% of life's time on this earth.   Stated differently, for
"80 percent of the time life has been on earth, it existed
solely as a single-celled microorganism" (Day 1984, 235).   Most
evolutionary processes appear to be largely irreversible.
       7.   That life created its own conditions for existence
reinforces the essential irreversibility of the evolutionary
process.   Technology is also essentially irreversible in that we
have adapted our lives to it and by having used certain resources
have probably closed the option of ever trodding the same path of
evolutionary change again.   Like it or not, if we humans are to
survive on this planet, it will be by carrying forward the
science and technology that we have.   As Rosa (1982, 37) puts it
for biological evolution, the building blocks of life, i.e.,
methane, ammonia, etc., are used up.   "Conditions are not the
same as when life arose the first time.   The niche for naive life
is closed and all life today is born of existing life."
Similarly, Day sees the process as "locked in step with the
earth's evolution, life passes only once . . . [for] once it had
come into being, life would go in only one direction--from simple
to complex."   One might add that all technology is born of
existing technology and the niche for naive technology is closed.

TOOLS, SIGNALS, AND ANIMAL BEHAVIOR
       We have thus far looked at various episodes in life's
history in an attempt to gain understanding of current
development processes and to raise questions about development
theories based upon biological argument.   The events we have been
discussing took place over the course of 2 billion or more years
from the probable origins of life to the development of life
forms with skeletons or hard exteriors about 600 million years
ago.   The episodes to which we now turn took place over the past
few million years.   These involve the tools, the language, and
the evolution of human beings.   Some of the tool using and
signaling behavior for animals other than humans may have origins
much earlier than a few million years ago, though all our
evidence on this behavior is from recent observation.   The human
part of our story begins about 4 million years ago; tool-using
humans or proto-humans began about 2.5 million years ago; and
language development probably began well within the past million
years.
       Animals of many species from insects to primates use tools,
just as they communicate with others of their species.   A
burrowing wasp disguises her nest by pounding the ground smooth
with a pebble hammer that she holds in her mouth.   Sea otters use
rocks as anvils to help them open shells.   Beavers, working in
groups, construct large and intricate dams (Oakley 1954, 1-37).
       The most striking observations of tool-using animals are

those made by Jane Goodall of East African chimpanzees. These
chimpanzees use tools for eating, grooming, and fighting. The
chimpanzees' tool use is unlike other species' because the
chimpanzees apparently are not using tools in the "stereo-typed
sequence of movements" observed in other animals, but rather the
young learn by observing their elders. A hungry animal picks up
or breaks off a stick or blade of grass and carries it some
distance to a termite hill. There, after knocking off the top of
the hill with his hand, the chimpanzee takes the stick and
dangles it down an exposed tunnel. He draws it out and picks off
the clinging termites for his dinner. Later, the same chimpanzee
may crumble and chew leaves into a spongy mass, which he uses to
sop water out of a stump for an after-dinner drink (Goodall 1963,
272-308; Goodall and Van Lawick 1965, 802-31; Lancaster 1968,
56-66). These actions clearly involve intelligence and
foresight. The rocks, the stick, and the sponges are used as
tools. Sometimes acquired before the end use is in sight, they
make the jobs of getting food or protecting the young easier.

The use of tools by some chimpanzees would seem to be more
fortuitous than essential. In any case, the tools are not
developmental in that they lead to further refinements or more
complex tools. However, it can no more be argued that these
animals have a technology simply because they have tools than it
can be argued that they have a language because they signal to
one another.

The vocal and nonvocal signaling systems of animals indicate
both feeling and information. These are not language but "innate
movements and sounds for expressing feelings . . . [and] innate
ways of reacting to these signals" (Lorenz 1962, 27-30). Bees
dance to indicate not only a new nectar find, but also its
location. The closer the nectar, the faster the dance. Any hive
member witnessing the dance knows exactly what course to pursue
(Hockett 1960, 89-96). A Siamese fighting fish identifies the
sex of another within the species by its reaction to the opening
movements of what may be a dance of love or a dance of death--
"not simply by seeing it but by watching the way in which it
responds to the severely ritualized, inherited, instinctive
movements of the dancer" (Lorenz 1962, 27-30).

Vocalizing animals call to each other in many ways:
defiance, warning, or greeting. The termite-fishing chimpanzee
has a range of calls that other chimps understand. Many birds
have complex codes that others in the species respond to even
when they have not heard them before. Such signals are automatic
and unemotional, innate and not learned. They do not rise from
the consciousness of the caller.

> All expressions of animal emotions, for instance, the
> "Kia" and "Kiaw" note of the jackdaw, are therefore not
> comparable to our spoken language, but only to those
> expressions such as yawning, wrinkling the brow and smiling,
> which are expressed unconsciously as innate actions and also
> understood by a corresponding inborn mechanism. The "words"

of the various animal "languages" are merely interjection
(Lorenz 1962, 89-90).

These signals are signals because their context is already
known.   They are restricted to particular situations.   Novelty,
if it exists, consists of a limited range of information such as
direction to food sources or location of dangers.   Beyond this,
these systems are closed.   A new experience cannot be shared
(Hockett 1960).

THE DYNAMICS OF LANGUAGE, TECHNOLOGY, AND HUMAN EVOLUTION
        Human language is open-ended; it is continuously able to
generate new combinations of ideas.   Any sentence can be a
combination of words never before used together, yet a group of
listeners can understand it.   Furthermore, language can be used
reflectively.   As linguists state it, only humans using language
can talk about talking:

        So, when we talk about a language--English, French,
Russian, Sanskrit--we refer to a process rather than to a
thing. . . . It is best to regard the language as a growing
corpus of words and structures which nobody can know
entirely but upon which anybody can draw at any time--a sort
of unlimited bank account.   It is not just the sum total of
what has been spoken and written; it is also what can be
spoken and written.   It is actual and potential.   In another
sense, it is a code always ready for individual acts of
encoding (Burgess 1965, 16-17).

Humans have the ability to use language reflexively because
we have a brain capable of self-examination.   As Henig (1983, 1)
says, "The human brain is the only organ on Earth that is aware
of itself."
        Taken alone, tools and signals are static.   Humans use them
and generate the dynamic processes of technology and language.
For the East African primate a stick has an end in view and its
usefulness essentially terminates there.   When humans have a tool
they find new uses for it, or they combine it with other tools to
serve new purposes.   The usefulness of a tool goes on and on.
        In the same way as language, technology is dynamic and
open-ended.   They are both subject to change and modification.
When we confront a new problem, we can use existing words to
create the sentences to describe it and discuss it, while we use
existing tools to create the means that will solve it.
        The dynamics of tool using and of open-ended language are a
function of the evolutionary process from which human beings
emerged.   During the course of this process, an interaction
occurred between tool using and biological evolution in a
feedback mechanism that provided a nonteleological direction to
change.   Currently there is some dispute as to the age and
significance of certain early skeletal remains, such as those of
the famed Lucy.   For the purpose of our analysis, we need not be

concerned about larger structure or the time frame of this transition. If indeed humans are 4 million or more years old, then tool using was not a critically differentiating causal factor. If the important transitions came later, on the order of about 2 million years, then the interaction with tool using was an important component of this critical differentiation from other primates. Even those who argue for the early origins recognize the importance that tool using began to play in the evolutionary process about 2 million years ago.

The feedback mechanism as described by Sherwood Washburn (1960, 63-75) is basically simple. As proto-hominids, also called hominoids, began using tools, there was a selective survival advantage in the group that had greater biological capacities for tool using. These characteristics are many and include the size of the area of the brain that controls the hand (in particular the thumb). As members of groups, or as groups within the larger population, that had greater capacity for tool using survived and bred, the population in general had a more favorable set of tool-using traits. Any random changes that favored tool using would be likely to survive and spread throughout succeeding generations. As these traits were intensified, the population's ability to make and use tools would improve. This greater physiological (which includes the brain and mental capacity) ability for tool using, combined with the nature of tool using itself, gave rise to new and improved tools. These improvements in tools and tool use then feed back to give further selective advantage to those members of the population with improved capabilities for tool using.

As this interactive or feedback process continues, hominoids evolve into hominids, and the process of tool using takes on the open-endedness of tool combining and becomes technology. Stressed throughout this book is the importance of understanding the context or ecology in which a technology operates. The preceeding illustrates an important component of technology, which we sometimes take for granted because of its universality, the physiological characteristics of the creators and operators of the technology.

The larger area of the human brain that controls the hand, controls a truly marvelous mechanism. Bernard Campbell (1982, 47) has a delightful description of the hand:

> Not needing our hands for support, we have been able to use them for more complicated and more creative tasks. With twenty-five joints and fifty-eight distinctly different motions, the human hand represents one of the most advanced mechanisms produced by nature. Imagine a single tool that can meet the demands of tasks as varied as gripping a tool, playing a violin, wringing out a towel, holding a pencil, gesturing, and—something we tend to forget—simply feeling. For, in addition to its ability to perform tasks, the hand is our prime organ of touch. In the dark or around corners, it substitutes for sight. In a way, the hand has an

advantage over the eye, because it is a sensory and a manipulative organ combined. It can explore the environment by means of touch, and then immediately do something about what it detects. It can, for instance, feel around on a forest floor for nuts and roots, seize them on contact, and pop them into the mouth; when your eyes read the end of this page, your hand can find the corner of the page and then turn it.

The hand itself may be a marvelous tool, but it is used to full value only when it manipulates still other tools. This capacity is a second-stage benefit of upright walking. With our erect posture, our hands are free; with hands free, we can use tools; with tools we can get food more easily and exploit the environment in other ways to ensure our survival. Humans are not the only animals that employ tools, but they are the only ones that do so to any great extent and with any consistency.

Campbell (p. 311) also suggests that the larger brain size enhanced the capability for attention span and memory. This greater capacity for memory would facilitate the cumulation of knowledge, including the knowledge of tool making and tool using. Individual and group memory would be the key component in the cumulation of knowledge and culture until this process gave rise to extra somatic means of cumulation in the form of writing.

The large human brain is "energetically expensive," and its evolution was possible only because of a generalized and energy-rich diet that resulted from simple technologies such as the digging stick to gain access to "energy-rich tubers" and from hunting (Lewin 1982, 540-41).

Stini (1980, 131) argues, "The human brain consumes about two-thirds of the circulating glucose and about 45 percent of the oxygen supply. This means that the human brain requires, on the average, from 100 to 145 grams of glucose per day. That translates into 400 to 600 kilocalories, or from a quarter to a third of the body's total requirement."

TECHNOLOGY AND HUMAN DEVELOPMENT

The "total life-way" of the tool-using human being created what C. Loring Brace (1967, 56) calls a "cultural ecological niche." This complex of humans and tool use is an extraordinarily successful adaptation. The dynamics of culture and the tool using part of this complex have allowed humans to spread across the globe and live in a vast array of climates and conditions without the necessity of biological evolution. This global dispersion, without speciation, has been taking place for well over 100,000 years and in a manner that is unique in mammal life. Based upon the "competitive exclusion principle" (i.e., "that no two organisms can occupy the same ecological niche"), Brace argues that "there has been only one hominid species at one time, and that the hominids of different time levels are linearly related." Evolution is sometimes thought of as the adaptation of the organism to the environment or changes in it. In a very real sense, technology involves the adaptation of the environment to the organism.

In an article surveying the latest studies in the early development of technology, Zvelebil (1984, 314) writes that "the analyses show that devoting time to the production of special tools may have made possible the replacement of Neanderthal man by modern man." The "making of tools takes time--time which could be spent on other activities, including the search for food. Benefits derived from the use of any range of tools must be carefully weighed against the time and effort needed to make them, and their manufacture and the search for raw materials must be organized so that they complement other activities required for subsistence."

Earlier it was noted that the large human brain and the enormous energy it requires is a highly inefficient evolutionary development unless there is a compensatory benefit. Similarly, the use of tools requires important amounts of time and energy of a group and is warranted only if its benefits exceed its costs. Obviously survival is a benefit that exceeds any achievable costs. A broader range of specialized tools created new resources and allowed for more effective exploitation of established ones. The competitive advantage of complex and specialized tools is operative where access to food resources is more difficult. This would refer, then, to the technologies that were necessary for humans to move from the tropics into cold climates. Since time is limited, the time-consuming activity of searching out food resources in more restrictive environments must be balanced by time-saving activities. Zvelebil finds that "foragers can increase the reliability and productivity of their subsistence strategy by using time-saving devices: by budgeting their time and by preparing in advance more specialized but also more complicated tools, designed for each of the number of tasks involved." It is consistent with the definition that we are using to call the more efficient budgeting of time organizational technology. More important, in the above quote is the concept of technology as time-saving. Clearly, to inaugurate a technological activity requires time to create the tools, as well as foresight. However, if the technology is warranted by the outcome, then the ongoing use of technology saves more time than its creations cost.

Activities that save significant amounts of time after initially requiring large investments of it emerge in agriculture. It is the large initial investment that leads some to argue that agriculture develops only if there is a food crisis as a result of population growth. Similarly, once a group is involved in agriculture, it takes another investment in time to go from hand tools to plows. But the time savings are enormous. One empirical study found that it takes about 4 times as long to prepare a field for planting with hand tools as compared to plows (Sanders, Parsons, and Santley 1979, 237). The critical building of an irrigation system requires an inordinate investment in time. Once built, sustaining it and extending the time costs are comparatively small, and the gains in increased output and in security and regularity of output are potentially quite large. Again, the point is that we have to distinguish between the costs in inaugurating a technology or technological system and the time cost in extending or utilizing a technology that is already a going concern.

The idea of time has been central to the economist's conception of capital and capital accumulation since the writings of Eugen V. Böhm-Bawerk (1930, 1957). To Böhm-Bawerk, industrial production was "roundabout" and capital accumulation was necessary to bridge the gap in time between the commencement of production and the consumption of output. The payment for capital was for the time that the accumulator had to forgo the consumption of it, so that it could be used in the productive process. In a biting attack against Böhm-Bawerk, Clarence Ayres (1944, 54) argued that "at any given moment the development of any industrial operation assumes industrial society as a going concern. This is indeed a sort of accumulation; it is the cumulative process of industrial technology." For those of us who are part of the "going concern" of an advanced scientific/technological/industrial (or post-industrial, as some would call it) society, our activities within it, be they growing food, producing goods, engaging in communication, making calculations, or traveling, are the fastest, most direct that humankind has ever known. The roundaboutness is the entire technological process, beginning with the earliest tools, since our modern technology would not exist without them, and the continuity of scientific and technological knowledge and the foresight to expand and apply it to human problem solving.

These distinctions apply to some of the issues involved in the selection of technology for development today. Accelerating development is comparable to initiating tool using. Complex, specialized technologies may be capital-intensive (i.e., roundabout or time-consuming to create), but they may also be the only technologies that can solve a particular problem or create land or other resources. Once underway, if the technologies are genuinely more efficient and not used merely because they are prestigious, they can be the most direct and effective, and possibly even capital-saving when measured in terms of output.

Thus far we have concentrated on the interaction between tools and human evolution. The very concept of evolution has generally implied continuous change, though there is now a school of evolutionary theorists (Steven Jay Gould and Steven M. Stanley among them) who believe in periods of rapid changes and long periods of species stability (Stanley 1981). Whether and to what extent humans are still evolving has become clearly secondary to the question of technological evolution and cultural change. Once primate evolution crossed certain thresholds, further biological change became increasingly less significant in explaining the development of human cultures, technology, and

civilization.   By the time of the emergence of <u>homo</u> <u>sapiens</u>
<u>sapien</u> (ca. 50,000 years ago?), further biological evolution was
of virtually no significance to cultural and technological
evolution.

Human culture has been called a superorganic phenomenon
to emphasize its distinctness and independence from organic
evolution (Kroeber 1917, 163-213).   Despite this belief in
separate realms of the biological and the cultural, some have
taken what they perceive to be the laws of biological evolution
and applied them to cultural evolution (White 1959; Sahlins and
Service 1960).

There are those who object to these conceptions of cultural
evolution, but nearly all anthropologists accept the idea that
cultural development and differentiation is not dependent upon
further human evolution and probably has not been for at least
the last 100,000 years or more.   In fact, Robert Lowie (1940, 3)
quoted E. B. Tylor on culture being "capabilities and habits
acquired by man as a member of society," and Lowie had culture
include "all these . . . in contrast to those numerous traits
acquired otherwise, namely by biological heredity."   However,
the development of culture and technology involves continuous
interaction with the biological realm in general and with
evolutionary change, in particular, of plants and animals.

One anthropologist, Walter Goldschmidt (1967, 110) has
defined <u>technology</u> as the "learned means by which man utilizes
the environment to satisfy his animal wants and cultural
desires."   It is then, in effect, a problem-solving process.   We
will use that as a preliminary definition of technology.   As
particular technologies are discussed throughout this chapter,
the definition will be refined, extended, and clarified.   Tool
use, like signaling, is widespread in the animal kingdom.   The
uniqueness to humans of technology is that it is productive,
open-ended, and therefore inherently developmental.

How long humans have been humans is a matter of definition
and of scientific debate.   The numbers can range from 500,000
years, if one uses evidence of language to several million years,
if one uses minimal physiological criteria.   Cohen (1977, 5)
refers to the "four million year history of Homo sapiens."   Most
of the larger figures refer to the formation of the family
Hominidae (i.e., hominids), and as a subgroup of the superfamily
Hominoidae (i.e., hominoids).   Many of the sources appear to use
the terms <u>hominid</u> and <u>human</u> interchangeably.   Further, for 99% of
his existence, man (the hominid) has been a hunter and gatherer
"tied to the seasons of vegetative food or movement of game."
With the exception of fire, he had "no power beyond that of his
body" (Spier 1970, 9; Fagan 1980, 148).   According to Lee and
DeVore (1968),

> Cultural Man has been on earth for some 2,000,000
> years; for over 99% of this period he has lived as a hunter
> gatherer.   Only in the last 10,000 years has man begun to
> domesticate plants and animals, to use metals and to harness
> energy sources other than the human body . . . .

Lee and DeVore estimate that above 80 billion people have lived out a life span, and of these, over 90% have lived as hunters and gatherers, compared to about 6% as agriculturalists, with the remaining few percent living in industrial societies. They conclude that "to date, the hunting way of life has been the most successful and persisting adaptation man has ever achieved."

Whatever number is used as a base, for the vast majority of the time that humans have had tools they consisted largely of basic stone tools. These define what we call the Paleolithic Era and the generally very short period of the Mesolithic. Humans are considered to have been tool users for at least 2.5 million years.

STONE TOOLS AND THE BEGINNINGS OF TECHNOLOGY

At one time archeologists referred to eoliths, or dawn stones (Braidwood 1964, 33). Eoliths were presumably stone tools that looked much like stones from streams or glacial creep. One author distinguishes between naturefacts and artifacts: nature-facts are "objects extracted from their natural setting and subsequently used without modification. . . . Artifacts are forms created by withdrawing materials from their natural setting and modifying them in trifling or remarkable ways" (Oswalt 1973, 14, 17-18). Most of what we observe used as tools by nonhumans are naturefacts. Naturefacts may have been the first tools used by humans, but virtually by definition it's next to impossible to verify or falsify this speculation. The tool evolution described below is artifactual in character because all involve ideas that are used to transform the material world; they are not merely extracted from it.

Andre Leroi-Gourham (1969, 21-25) describes a series of stages in human tool making. The first two stages of this taxonomic scheme consist of core stone tools (pebble tools, as they are called) and flaked stone tools. Both types are made and not merely found (Braidwood 1964, 33). Leroi-Gourham describes the third stage in his classification as "a major step in the history of the human race . . . for it served as the foundation for the conditions of technological development until the appearance of metallurgy." In this stage, the techniques of the previous two stages are combined in that a stone core is first worked and prepared and then the desired tool is flaked off from it. "Production begins with a core, or nucleus of raw materials, as before," Leroi-Gourham writes, "and its end result is a kind of surface that is dissymmetrical in thickness and has the shape of a tortoise's shell." In the fourth stage these techniques are used to create a wide range of tools, i.e., scrapers, grovers, drills, and blades.

In the third and fourth stages, tool using gives rise to technology (as it has been defined here). Tool making becomes combinational and more rapidly cumulative. The tool-making techniques "require a well developed feeling for the material, some preconceptions of the desired result, and a manual skill that would not be deduced from an examination of the maker's cranium." Leroi-Gourham's observation here brings out two other

characteristics of technology: all tools and technologies
involve ideas or "preconceptions," and even seemingly simple
stone tools involve a complex interaction between skills, ideas,
and materials.   It is the replication of preconceptions in a
series of tools that allows archeologists to define tool
traditions (Braidwood 1964, 40).

The "mastery over materials" characterized what is called
the upper Paleolithic.   New materials were used in tool making
(Oakley 1954, 32; Fagan 1980, 101).   Kenneth Oakley writes,
"Artifacts of complicated forms were wrought in bone antler and
ivory by a combination of sawing, splitting, grinding and
polishing.   By now, tools were not only used to make implements
in the sense of end products, such as meat knives or spears, but
many tools were made which were tool-making tools.   This is good
evidence that the hunter-craftsman was showing considerably
greater foresight, and no longer worked merely to satisfy
immediate ends."   Just as earlier different ways of making tools
were combined to make better tools, new tools were created as
composite tools or as a combination of existing tools.   The use
of those new tools involved the application of new principles of
mechanics and power.   According to Oakley, "Spears were launched
with throwers which, working on the lever principle, increase the
effective propelling power of a man's arm."   The invention of the
bow brought additional capabilities.   Again, from Oakley we learn
that the bow was the first means of concentrating muscular energy
for the propulsion of an arrow, but it was soon discovered that
it also provided a means of twirling a stick, and this led to the
invention of the rotary drill."

A greater variety of tools allowed for more intensive
exploitation of new environments.   Use of fire and of clothes
made from skins allowed humans to move out of the tropics into
the colder climates of the late Pleistocene.   This northward
movement began earlier with the transition from core tools to
flake industries (Clark and Piggott 1965, 59-60).   A variety of
materials became resources for human use.   "Bone and ivory
bodkins, bone needles with eyes, belt-fasteners, and, rarely,
even buttons have been found in Upper Paleolithic sites," writes
Oakley.   "Carved representations of clothed figures . . . show
that these hunters wore sewn skin garments with fitting sleeves
and trousers."   This clothing improved efficiency and the
hunter's ability to survive very cold winters.

A more intensive exploitation of the environment allows for
a population buildup and the creation of more permanent
settlement patterns.   Mesolithic industries included a variety of
tools for gathering food.   One of the defining characteristics of
this period is the abundance of microliths, which could be used
in a variety of composite tools for hunting and gathering.
Settled communities are a more efficient means for assembling,
cumulating, and diffusing technology.

Gradually we begin to see the build-up toward the first
basic change in human life.   This change amounted to a

revolution just as important as the Industrial Revolution. In it men first learned to domesticate plants and animals. They began producing food instead of simply gathering or collecting it (Braidwood 1964, 97).

As with most technological transitions, many of the tools used for gathering were later used in agriculture.

Though the transition from hunting and gathering seems in retrospect to have been gradual, from the perspective of that which preceded, the change was explosive. There was at least a million years (and possibly 4 million years) of human history until the period that we call the upper Paleolithic began about 35,000 to 40,000 years ago. Using the figure of a million years for human history, 98% of the time humans have been on Earth, we have been hunters and gatherers. About 20,000 years ago a shift began from a hunting and gathering lifestyle to a more specialized economy. This brought the use of storage pits and of ground stone tools used to crush pigments and tough grass seeds (Fagan 1980, 153).

About 10,000 years ago the world's population of about 10 million people (possibly as high as 15 million) was almost completely dependent upon hunting and gathering. By 2,000 years ago the overwhelming majority of people lived by farming (Cohen 1977, 5; Fagan 1980, 149). As Kenneth Oakley put it, homo sapiens "ceased to be a rare species." There were then between 100 and 200 million people on Earth. In 1500 more years, the population grew to about 350 million and agriculture spread to most areas manageable by the technology of the time. Harlan (1968, 6) writes that "by the beginning of the 20th century (A.D. 1900) when modern ethnographers had begun their observations, the world population had jumped to about 1.6 billion and the hunters and gatherers decreased to less than 0.001%." He calls the hunter-gatherer population "heading toward extinction."

TECHNOLOGY AND AGRICULTURAL ORIGINS

Agriculture was made possible by advancements in technology and concentration of technology. Similarly, agriculture allows for the further increase in population (Wenke 1980, 268). Some archeologists argue that it was overpopulation relative to hunting and gathering technology that forced the near universal transition to agriculture (Cohen 1977). However, these explanations seek the "why" of domestication. How it happened still depends upon technology. Not only does the direct process of agriculture (i.e., tending the fields) require a more sedentary lifestyle, but all its consequences further restrict mobility. Properly done, agriculture requires a wider range of tools than would be possible to carry about. "Moreover, storage technology is required," Wenke (p. 275) writes, for "where could one go with a metric ton or so of clean wheat seed, no matter how nutritious?"

Braidwood (1964, 113) suggests that a "man who spends his whole life following animals just to kill them to eat, or moving from one patch to another, is himself really living just like an animal." Clark and Piggott (p. 130) say hunting and gathering set "narrow boundaries to development." Wenke (p. 266) argues that "the correspondence between agriculture . . . and civilization is absolute".

It has long been noted that some of the most significant human foodstuffs, namely the grains, are part of a larger category of plants that we call weeds. This quality of "weediness" is the requirement that they must grow in aerated soils. As human technology became more specialized and complex, allowing more intensive exploitation of the environment, then small sedentary communities could form. The very activity of creating a site, building shelter, etc., disrupted the habitat and frequently broke up the ground, aerating the soil. Many of the traditional theories on the domestication of grains (and other cultigens) relied on humans dropping or passing undigested seeds they had gathered and then later observing their growth in the settlement area. The people would eventually begin to artificially aerate larger areas in which they would then broadcast the seeds.

Hawkes (1983, 7) clearly defines the concept of "weediness" both in a horticultural and in an ecological sense.

> One of the characteristics of cultivated plants and their wild ancestors is an inability to compete successfully with natural climax vegetation, and a marked preference for open, disturbed, or ruderal habitats with bare soil and a minimum of competition with other species. They are thus spoken of as ecological weeds rather than horticultural weeds ... The ecological definition of a weed is a plant of secondary successions, that is, one establishing itself and growing quickly on bare soil, almost regardless of the nutritional status of that soil, but dying out rapidly when the soil becomes covered by perennial grasses, herbs, shrubs, and trees. Often, but not always, weeds require high levels of nitrogen in the soil.

To Hawkes (p. 30), many of the plants that became human cultigens were "once nature's misfits":

> Before the advent of man, they must have lived a precarious existence on river banks, sand bars, and game trails, and in areas where landslides occurred or animals wandered or bedded down. It is likely that they were chiefly adapted to regions where poor, thin soil dried out quickly at the end of the rainy season and so prevented the establishment of trees, bushes, and perennial grasses with which our crop ancestors were unable to compete.

Hawkes (p. 32) argues that the severe environments in which weeds could survive gave rise to characteristics such as large seeds that "germinate and grow quickly and provide a reserve for the dry season." This was a form of "pre-adaptation" to agriculture. Rather than humans being responsible for domesticating the plants, Hawkes suggests that the plants may have invaded the open areas created by human settlements. It is interesting that many of the prevailing theories for the domestication of some animals (such as dogs) have the animals domesticating themselves.

There was an interaction between human tools and plant evolution that was similar to the earlier interaction between tools and human interaction. Weeds propagate by dispersing their seeds widely. Once humans enter into the process, seed dispersal is no longer as important for perpetuation of the plant type. Neither was seed retention of particular benefit to gatherers or "under cultivation if crops are harvested by hand-stripping. However, once the harvesting of crops by sickle was introduced, the seed retention character became crucial" (Evans 1980, 389). Interaction with the sickle gave direction to the plant evolutionary process towards greater seed retention, so that one of the characteristics of most domesticates is the inability to propagate (let alone survive) without the intervention of humans and tools.

Most of the scenarios that arise from the theories of domestication of plants and animals have a fortuitous element to them. Following Hawkes' thesis of an invasion of human settlements by certain plants, the operative human intelligence was more in terms of discovery than invention. Throughout the evolution of science and technology, accidental discovery plays a role. Recognition of this fact in no way diminishes the quality of the human intellect in the invention or discovery process. Humans operating intelligently in problem-solving processes can bring about larger transformations beyond what is possible for them to know at the time. Clearly, the early tool-using proto-hominids knew they had a competitive advantage in making and using tools (why else do it?) but not that they were bringing about an evolutionary, biological change. This longer transformation is what the sociologist Robert K. Merton (1949, 66) called the "unintended consequences . . . of a given practice."

Many early simpler tools were made on the spot with a fairly short-term end in view. Later, as tool assemblages became more specialized and complex, a longer planning horizon emerged. In colder climates, for example, humans made many of their tools in winter for later use in summer. With domestication, the planning horizon extended even further, as people planted a valuable foodstuff in order to have more at a later date. Specialists may debate the extent to which early cultivators and those engaged in animal husbandry recognized their role in the breeding of plants and animals. Whatever one's position is, it is difficult to argue against the thesis that the farther along in the process, the more likely is the awareness of long-range evolutionary potentials. We learn from doing and from observing what we have done. Though chance continues to be a factor in science

and technology today (and probably always will be), we are
increasingly able, through understanding the past processes of
change, to plan our research and development in providing for
more effective science and technology policies.

TECHNOLOGY AND MATERIALS TRANSFORMATION

The development of agriculture requires methods of storage.
Food storage is a form of environmental modification that helps
to smooth out the irregularity of food supply over the seasons.
Humans may have evolved in areas of yearround food supply;
agriculture tended to evolve where there was a discontinuity in
food supply (such as in areas with dry seasons). Food storage
and farming practices would together interact to bring about
agriculture. Farming is primarily planting; the development of
agriculture involved genetic change in the plants themselves.

Many foodstuffs require processing and cooking for digesti-
bility. Pottery is useful in both these endeavors, though
not essential. Pottery was independently developed in many
locations. It is described by one author as a hallmark of the
Neolithic period: "Pottery or ceramic ware is one of the first
synthetic materials created by man. It owes its existence to the
irreversible change brought about when clays are heated to drive
off the water" (Spier 1970, 41). The use of stone, wood, bone,
vegetable fibers, and a host of other items involves changing the
shape of the material to use it as a tool. The use of fire
allows humans to transform the internal structure and composition
of materials to make them more useful to themselves. Previously,
resources were found; now they are also made. Pottery making
reflects some of the open-endedness of technology. It also opens
the possibilities for continued improvements in the materials
themselves. Ceramics today have many traditional and new,
sophisticated uses, from cooking to spacecraft, and research
continues on material improvements and new uses.

Materials transformation has become the basis for subsequent
periods for our naming of stages of human technological
achievement, i.e., the Bronze Age, the Iron Age. Though metals
were probably first used as found, in comparatively pure nodules,
such as copper, the smelting of ores was soon necessary if
their use was to be sustained. Transforming ores into metals
is transubstantiation. We know today the nature of the transformation process; is it any wonder that those closer to its
origins thought they also could transmute "base metals" into
gold? As with the development of ceramics and pottery, the
development of metals inaugurated an open-ended process that
allowed for the creation of new forms, continued improvements in
the quality of the materials used, and the opportunity to turn an
even greater part of the universe's materials into resources.
Those materials, furthermore, can be combined with a variety of
others to create even more useful materials. The quality of the
stone strongly affects the quality of stone tools, but the
quality of the ores has progressively less to do with the quality
of the metal tools than do the science and technology of
production.

Raw materials and other aspects of the environment become resources when humans acquire the ability, or the science and technology, to exploit them. Certain rocks become resources when the knowledge and skills to make stone tools evolves. Roots become food resources when stone tools to dig into the hard ground are available. Large game become resources when hunting tools (such as bows and arrows) and social organizations emerge. Certain seeds become resources as humans develop tools and techniques for more efficient intensive harvesting. The resource character of these seeds is enhanced as the evolutionary changes of domestication increase their yield. Various types and kinds of ores, too, become resources--natural resources, as we unfortunately call them--as we continue more efficiently to learn to extract from them what we wish to use. Technology, then, creates resources.

TECHNOLOGY, DIFFUSION, AND THE HUMAN FUTURE
Many of the important inventions of human history in the domestication of plants and animals appear to have taken place independently in many different places. However often the same item may have been invented, no one can assert that all peoples developed all of their technology. This straw man is raised simply to state the point that throughout human history people have borrowed technology from their neighbors. Consequently, the diffusion of technology (or technology transfer) is essentially as old as technology itself. Furthermore, every tool involves a human skill. Skills involve people doing things, which is behavior. Some tools, such as those used in paleolithic hunting of large animals, involved group behavior and therefore social organization. One need not have a determinist model of technology and social organization to recognize that, though there may be a range of possible behaviors and social organization, tool using nevertheless implies in a broad sense forms of behavior and organization. People also developed belief systems that include ideas about technology and social organization.
With the growth of sedentary communities, there is a parallel growth in technologies peculiarly adapted to the particular environment of the group. Of course, from the beginning people developed or adapted technologies to solve their problems. The forest or the savanna, the tropics or the colder climates, each gave rise to different technology or variations on the same technology (depending on one's definition). Tools, then, can be related to the environments in which they are used (DeGregori 1969, 83-125).
After 4 million years of hominid history and 2.5 million years of tool using, have we humans reached the limits of our environment beyond which even further evolution or technological change cannot carry us? The answer is emphatically no! The fundamental purpose of most of the succeeding chapters is to explore the ways that further scientific and technological change can continue to expand human horizons and opportunities. Cohen (1977) and others argue that humans reached the limits of their environment 20,000 to 40,000 years ago, as defined by hunting and

gathering technology. The food crisis, as he calls it, forced humans to devise new means of agriculture to solve the problems wrought by overpopulation, relative to a technology of food supply. Another author, Ester Boserup (1965), has argued that overpopulation relative to food supply has been a continuing force for agricultural innovation throughout history.

At the time hunting and gathering supposedly had reached environmental limits, the population is estimated to have reached 10 million to 15 million. Perhaps this was not in fact the actual limit of this technology. Assuming that hunters and gatherers could harvest and utilize 100% of the natural environment's production, the world's ecosystem could support possibly 200 million people. This figure is 5% of the current world population. Even if we don't accept the 10 million to 15 million population as being the hunting and gathering limit, it was probably quite close to the exploitable portion of the natural ecosystem. Clearly, we are by irreversible necessity committed to a post-agricultural revolution technology. Further, it demonstrates the way in which changes in technology can dramatically change the economically exploitable limits of our environment.

Are the soil, water, sunshine, air, and all materials that humans use on this globe to create and sustain life sufficient to the demands that must be made if human life is to continue? In the chapters on catastrophism and the limits to growth, we will actually define these limits and indicate that we are far from reaching them. That does not mean that we will automatically come up with the advances in science and technology needed to transform the earth and improve the life of a growing population. As with technological change throughout human history, advances are not automatic but require thoughtful, intelligent action directed toward problem solving. As we look back at the history of human problem solving, we get a better understanding of the process so that we are better able to act intelligently and expand our time horizons to predict and account for longer and longer term consequences of current actions. We have taken the first hesitant steps into space. Long before we have exhausted terrestrial life-sustaining resources we will have the capability of exploiting the virtually unlimited resources of our planetary system and beyond. This reality, though, is far beyond the more urgent and immediate needs of economic development and the elimination of poverty that face us and are the subject of this book.

We have spoken of life, language, and technology as being dynamically open systems. In technology, the dynamism comes from ideas. Douglas Hofstader (1983, 14-21) compares language and ideas to dynamic evolutionary life processes. He speaks of "virus-like sentences and self-replicating structures." Like life forms, successful ideas gain an ecological niche in "idea space." Quoting R. W. Sperry, he finds that "ideas cause ideas and help evolve new ideas." Just as there is a biosphere, there is also an ideosphere. And ideas have "spreading power," what we in our context might call technology transfer. Just as life and other self-replicating systems can be understood in terms of their ability to survive, so also do ideas have a performance value [that] depends upon the change it brings to the behavior of the person or group that adopts it. Technologies as educational processes must be sustained by the performance value of the ideas embodied in them. Technology must allow people to change their behavior and improve the quality of their life. Technology as an idea system, like the life system from which it is derived and compared, must have the survival value of sustainability in and through the life processes of the humans who use it. Technology and technology transfer then have empirical, testable, measurable consequences.

As Lee and DeVore note (1968, 3) stratigraphically the origin of agriculture and the present "will appear as essentially simultaneous. If we fail, then, it will be a tragedy we fail, then, it will be a tragedy, not only for all those alive, but it will be a failure of civilization itself and an implicit argument that human life beyond hunting and gathering is inherently unstable." Tool using, as C. Loring Brace argues, has created for humans a unique ecological niche. For 125,000 years, we have spread across the globe into every known climate (and now into space) without biological adaptation of the species (Brace 1967, 56-57). Technology has been our means of adaptation and survival, and it can continue to be so. As with most human endeavors, the choice between success and failure is ours.

CHAPTER 2
Technology and the Historical Process

The mainstream of economics has long considered questions of technology as subsidiary to those of saving and capital formation. However, in the past two decades, the issues of technology transfer have become a major concern for those interested in the problems of economic development in less developed countries. Economists and others are now writing about scale of technology, appropriate technology, transfer of technology, and so on. This outpouring appears to consider technological diffusion as largely a post-World War II phenomenon. Of course, such is not the case.

Throughout human history technological diffusion has been a regular and important element in the evolution of a people's technology. The populations of all cultures and places use a technology, although they, or their ancestors, originated only a small part. Technological borrowing incorporates the creativity of the rest of the world. The process of adaptation in technological borrowing is itself a form of inventive activity. Yet, little of the writing on technology transfer reflects any attempt to gain understanding from prior successes (and failures) in borrowing and using exotic technology.

This chapter argues that the history of technology is replete with insights that yield theories directly applicable to the current pressing problems of development. Using this historical base, the nature of technology is defined in a way that generates useful and practical concepts for thinking about development and for carrying out programs and policies using technologies to raise the levels of living for the world's population. Toward the end, the chapter suggests that some views of technology transfer are deficient in historical understanding and, if carried through, have the potential for curtailing the long-term processes of development.

Technologies can transfer by many different means. People can borrow and adapt technology from their neighbors. Migrating or conquering people can carry it to new areas. People tend to carry much of their cultural baggage with them, which includes their social organizations and belief systems. Settled communities that more intensely exploit their environment give rise to larger population concentrations. Larger populations increase

the likelihood of larger armies. Advances in technology, such as
metallurgy, create the possibility for better weapons. Larger
armies and better weapons facilitate conquests. By no means is
it true that people with superior technology for economic
exploitation are always the military victors. Whoever the
victors are, there has been a relentless conquest by agricultural
technology over hunting and gathering technologies. Superior
technologies tend to conquer, even when the carriers of it lose
the particular war.

The diffusion of technology, along with beliefs and social
organizations, can create problems. Techniques in agriculture,
suited to one environment, for example, may not be operational
for another. The basic problem-solving character of technology
is not always fully understood. Misunderstandings occur over and
over in human history. The Romans brought Mediterranean
agriculture north of the Alps, complete with latifundia, a Roman
style of life (Brown 1967, 336-37; DeGregori 1974, 5-6).
Europeans brought their agricultural practices to North America.
In both instances, when the technology failed to perform as it
did in the homeland, the environment was deemed to be inherently
and eternally inferior for agriculture. Subsequently, however,
with proper adaptation of alien technology and its combination
with indigenous technology, both areas--Europe north of the
Alps and North America--became among the most agriculturally
productive in the world. Europeans also sought to diffuse their
technologies to areas of colonial conquest in Africa, Asia, and
South America. Again the environment was blamed for any
shortcomings in performance or output. Subsequent refutations of
these interpretations have in recent years gone to the other
extreme of blaming some entity called "technology" or "Western
technology."

Thus far we have concentrated on peoples using technology or
using the wrong technology. We have not discussed the institutional factors that cause people not to use a technology. One
institutional factor in not using a technology is that people do
not want to use it. It is not a question of whether a new
technology will solve a particular problem better, but it is just
that the belief system forbids it. Thus, in the United States
today there are groups that will not use motor vehicles and
others that will not use medicine or allow surgery. Such groups
are replicated through history and around the globe. The
efficacy of the given technology is irrelevant; their religion
forbids its use. There is another category of not using
technology that is far larger. As we have noted, the use of
technology involves the creation of social institutions that
define and validate its use. The very complex ideas and beliefs
involved in using a technology in time can be the basis for
denying the superior efficacy of a new technology. The argument
has been made that advances in technology and in science tend to
develop on the periphery of civilization where these institu-
tionalized idea systems are less well established and therefore
less able to resist pragmatic adjustments and changes (DeGregori
1969, 1974). Under these circumstances technology transfer to

new areas frequently not only allows the recipient to catch up,
but also allows the borrower to use it as a basis for surpassing
the lender.

COSTS OF MILITARY MYOPIA
     Though this book is concerned almost exclusively with the
technology of production of economic goods, the technology of
warfare can serve as an illustration of institutional resistance
to technological change. Whatever the conservative impulses are
in a culture, to lose a war is potentially to lose everything.
Consequently, in virtually all instances, people would not
knowingly preserve a military technology that would bring them
defeat. If we find repeated instances of such behavior, then the
likely answer must lie in an institutionally conditioned
blindness to the possibilities of new technology. A cursory
view of some selected aspects of Western military technology
will illustrate this interrelationship between technology and
institutional myopia.
     In the early phases of Middle Eastern civilization, military
supremacy was predicated upon bronze, then iron, then chariot
warfare, and later, the phalanx with iron spears and shields.
Bronze and chariot warfare limited fighting primarily to an elite
(since bronze was of limited availability and chariots and horses
were expensive), while iron was more widespread and cheaper and
favored a highly organized yeomanry, as in Classical Greece.
When the Athenian power base became heavily maritime, a new
social importance accrued to the urban lower classes who manned
the oars. Not only did these changes in technology bring social
changes, but they also left in their wake a series of defeated
peoples who could not respond to change.
     The foot soldier in the form of the Roman legionnaire with
his sword and shield was challenged by light horse cavalry during
the Gothic and Germanic conquest. The issue was not clearly
decided until the eighth or ninth century, with the development
of the stirrup, horseshoe, and heavy shock cavalry. To some
observers, heavy cavalry and the three-field system were
primarily responsible for the development of feudalism. Yet in
1066, two centuries after the development of shock cavalry,
Harold's troops rode into the Battle of Hastings on stirruped
horses, dismounted, and were defeated by the partially mounted
Normans (White 1962, 36-37). For nearly 3 more centuries,
stirruped cavalry was the most efficient form of warfare;
inculcated into European social consciousness were attitudes
toward the horse and nobility that persisted until very recent
times.
     In the fourteenth century the English longbowmen and the
Swiss pikemen drove heavy cavalry from the field, though
the latter continued to appear and to suffer defeats for at
least another century. In the nineteenth century a series of
inventions--the repeater rifle, the Gatling Gun, and barbed
wire--made cavalry clearly obsolete as a battlefield instrument,
a judgment that was confirmed by the trench warfare of World War

I.   In the same century the 2 most famous cavalry charges,
Balaclava and Sedan, were colossal failures.
     Despite all of this, in the pre-World War II period the
major powers (except the two losers, Germany and the Soviet
Union) were preparing to fight the next war on horseback, and for
the Poles it was even more tragic, as they sent a unit of lancers
to meet the German invaders.   There was no lack of empirical
evidence that the internal combustion engine (along with the
other previously mentioned inventions) had doomed the horse; the
brief tank warfare in World War I, the many interwar years
manuevers, the German tanks in Spain, and the Soviet tanks in
Outer Mongolia had all demonstrated this point.   The retention of
the horse and the near national catastrophes that it created were
the results of sentiments of nobility, honor, and chivalry that
harked back deep into the Middle Ages.   In describing similar
attitudes in the British Navy, B. H. Liddell Hart (1965, 326)
likened an Admiral's attitude toward his ship to that of a bishop
to his cathedral.   Such sentiments may win the praise of the
faithful, but they do not defend a nation.
     The classic 1930s statement confusing the spiritual value of
the horse with its military technical efficacy was made by
British Field Marshall Haig: "As time goes on you will find as
much use for the horse--the well bred horse--as you ever have
done in the past" (Hart, 100).   This digression should not lead
us to believe that this attitude toward the horse as a military
instrument was inherent in the horse itself.   Exactly the
opposite attitude and pattern emerged in China.   There the horse
was identified with "barbarians."   There, for over a millenium,
Chinese leaders knew that they must raise and use the horse
as a battlefield instrument if they were to succeed in defense
against mounted invaders.   Try as they might, they could never
disassociate the horse from the "barbarians" who brought it and
developed that close man-animal relationship necessary for its
effective use.

CULTURE AS A KEY FACTOR
     The vast majority of the technology of any people was
developed by others.   We are all or have been borrowers of
technology.   Technology transfer is a constant facet of human
history.   The study of some of the more dramatic of these
transfers, although not providing all the answers for present
problems, can nevertheless give insight into some of the most
important characteristics of technological diffusion.
     Northern Europe and, later, North America built industrial
economies based on science and technologies that originated in
other areas.   The religion, science, and technology of medieval
Europe had been historically alien to that region.   Their origins
were in the greater Middle East.   For several thousand years
there was a continuous, fruitful interchange of tools and ideas
in the area that included the Indian subcontinent on the east and
stretched to and later included the Hellenic, Hellenistic, and
Roman peoples in the west.   This civilization spread across the

countries on the shores of the Mediterranean Sea. Politically it
became dominated by Rome.

Small elements of these great civilizations were filtering
northward into Europe. While there were great civilizations in
India, in the Middle East, and in the eastern Mediterranean,
northern Europe was in comparative darkness. With Roman
conquests north of the Alps came a significant increase in the
northward movement of Mediterranean technology. This movement is
interesting and significant because virtually every error made in
attempting technology transfer during the periods of and after
European colonialism was a replication of those made during Roman
colonialism.

Many Roman writers thought that northern Europe was essentially uninhabitable. Among other reasons, it was too cold.
Rome had taken Mediterranean agricultural practices, crops,
architecture, and styles of life and had attempted to transplant
them in a different climate and geographical environment without
modification. No wonder, then, there was an apparent "failure of
technology." In fact, it was neither the inadequacies of the
peoples or environment of northern Europe nor was it a failure of
technology. The fault was in the character of the diffusion
process.

The diffusion of science and technology to North America is
comparable to the flow from Rome into northern Europe. In the
eighteenth century, Europeans debated the habitability of North
America. Though they did not question the physical heritage of
their kith and kin who were conquering the continent, some argued
that the climate stunted the growth of humans, other animals, and
plants. It is clear from subsequent history that northern Europe
and North America had great potential for agriculture and
industry, provided the right technology was used.

The key to the development of both areas was the shifting of
decisionmaking power from alien colonial authorities. The new
technology that evolved was a synthesis of indigenous technology,
exogenous technology that was modified and adapted to solve the
particular problems of the new environment, and, in some
instances, a direct borrowing of exogenous technologies that
needed no adaptation. What is important is that, in this
synthesis, the people of these areas made the technology part of
their culture. Tools and technologies are more than mere gadgets
or physical instruments; they are the embodiment of ideas. If
these ideas become a vital part of a culture, then there is more
than an assemblage of tools--there is a continuous, dynamic,
evolving process of change, technological borrowing, and
development. Technology evolves as the combination of existing
tools and technology. The more tools, the more possibilities for
combinations and acceleration of development.

The evolution of printing by movable type in fifteenth
century Europe is an example of technological change resulting
from a synthesis of indigenous traditions and adaptation of
exogenous technology. It also demonstrates the worldwide nature
of technology, as most of the "indigenous" elements of European

printing were the products of earlier diffusion of technology from other areas. Printing by movable type involves printing, papermaking, metal working, a phonetic alphabet, and an ink that adheres to typeface and to paper.

Evidence of printing can be found as early as the seventh century in Korea and the ninth century in Central Asia. There may have been earlier origins related to the practice of taking rubbings from stone in China. Paper derived from the felting process of making tents and garments in Central Asia and the use of these techniques to make rice paper in China. Papermaking diffused into Persia in the ninth century and into Italy in the eleventh century.

Printing by movable type is superior to other forms of printing if one has a phonetic alphabet. The phonetic alphabet is an instance of preadaptation. It is a trait that was developed for its own purposes--writing--that was not overwhelmingly superior to other competitive techniques used for that purpose. It was, however, almost uniquely adapted to an invention that did not occur until over 2 thousand years after its beginning. Preadaptation is a phenomenon of both biological and technological revolution.

The phonetic alphabet was part of Europe's heritage from the Middle East that was brought north of the Alps by the Roman conquest. Presses were independently invented in many different parts of the world. The European press descended from the olive presses of the Grecian world. Metalworking had come to Europe from the Middle East 2000 years previously and had taken root and developed to a high art. Ink from linseed oil was a purely indigenous invention derived from the flax used in textiles and was used in a rapidly growing tradition of oil painting.

Printing by movable type was but one of the many European developments in science and technology that were largely derived from Asia and/or the Arab world. Many of the important terms in European science and mathematics were derived from Arabic, reflecting the Islamic influence and origins. Europe was fortunate in being a crossroads for the movement of tools and ideas. Each of these was absorbed into European culture, becoming part of its tradition and was the basis for continued European development that led to the scientific and industrial revolutions.

The diffusion into Europe of alien ideas and technologies was an important element in the intellectual upheaval and the cultural change in Rennaissance Europe. One might even be so bold as to say that the change was revolutionary. Certainly, what was to follow in the next centuries was deemed after the fact the Scientific Revolution and the Industrial Revolution. Europe (and many of the areas where the population is predominantly of European extraction) has been experiencing 5 centuries of fairly continuous scientific and technological change and cultural transformation. Yet, even in the most industrially advanced European countries large pockets of traditional culture, such as that of the peasantry of France,

persist even with the intrusion of the automobile and television.
Who would argue that English or French cultures are any less
English or French today than they were 5 centuries ago, just
because they have undergone such enormous change? Industrialization has made France and England alike in many ways, but it has
not obliterated those distinctive features of these cultures
which are a source of pride and identity for its members and
constitute a vital part of the heritage of all humans.
     Rather than being destructive, technological change in
Europe has stimulated its arts and culture. Printing by movable
type was, as we have noted, developed in Europe in the middle of
the fifteenth century. Before the end of that century, more
books were printed in Europe than had been printed in the history
of the human race. "One calculation puts the total of incunabula
(books printed before 1500) at 20 million" at a time in which
Europe had "perhaps 70 million inhabitants." In the next
century, the sixteenth, it is estimated that between 140,000 to
200,000 editions were published, totaling 140 million to 200
million books (Braudel 1973, 298). This is more than just a
quantitative distinction, as great written works of many cultures
were being printed and reprinted in translation. Printing was
the key factor in the rapid growth of literacy, in the improvement of communication among scientists, and therefore it was
important in the growth of science and in the development of
literary forms. In fact, one could argue that the scientific and
technological revolutions have enhanced all of the arts, both in
their creations and in our ability to appreciate them.

LESSONS FOR TODAY
     The European experience in technological and economic
development provides a useful example in current attempts to
foster economic growth and development. Europeans, when
colonizing other parts of the world, made the same mistakes that
the Romans made when they colonized northern Europe. They
attempted to transplant their technology in agriculture, in
industry, and in education to their colonies without modification
to fit the cultural and environmental conditions of the area.
Though there were also many successes in technology transfer as a
result of culture contact between two peoples, many of these
schemes failed. The climate and the peoples of the colonies were
blamed. Just as Northern Europe was considered too cold by
Romans to be inhabited in a civilized manner, the tropics, it
was argued, were too hot and humid for long-term economic
development.
     It is ironic that these tropical and subtropical climates
that were considered unfit for agriculture by European
colonialists were the locations for the first domestication of
most of human agricultural foodstuffs. These climates are still
the poorer areas of the world and constitute the larger part of
what is called the third world.

     Agriculture began in what is now called the Third

World. The ancestral homes of the world's most important crops are in Third World countries: rice, wheat, maize, potato, cassava, sweet potato, sugarcane, soybean, pulses, numerous vegetables, most fruit trees, cotton and other fibres, many forage grasses and legumes, and numerous forest trees including most hardwood species (Swaminathan 1983, 553).

The failures of many projects of technology transfer in the colonial and post-colonial period have given rise to charges that technology has failed. Voices are raised on behalf of appropriate technology, or intermediate technology, or indigenous technology. Technology is a set of tools, machines, and ideas; it is a problem-solving process. When a set of tools designed to solve one set of problems is applied to different problems, it is understandable if those tools do not always work. It does not mean the technology failed, merely that the technology was incorrectly used.

The developed countries' experience with alien technology is not, however, a complete guide to the dilemmas of less developed countries. Countries like England and the United States have had a couple of centuries to industrialize and absorb the changes that come in its wake. The less developed countries do not choose and in reality cannot afford the luxury of this slow rate of economic growth. The pace of change today can be destructive without giving cultures the opportunity for creative, constructive response. The pace alone is threatening. In addition, the many facets of modern communication, such as films, radio, television, magazines, and consumer goods, bring slickly packaged chunks of exotic culture. These influences limit or at least make infinitely more difficult the task of leadership in providing responses to change that preserve and sustain local cultures. There is insufficient ground for despair or pessimism. Technology is not inherently destructive, as many claim. Quite the contrary is true. The challenge to sensitive, intelligent leaders and to the population of less developed countries is enormous. Exotic influences, though, are not going to evaporate without a trace, and the possibilities for personal and cultural enhancement are virtually limitless.

For those in many parts of the world, "Western" science and technology may seem alien, but in reality it is their own ideas and tools coming back home to them in another form. For it was Greek, Arab, Indian, and Asian science and technology that formed the basis for European development. In a larger sense, exotic technology is coming home to all people, for it is the heritage of all mankind. Science and technology are both universal and particular. They are universal, for example, in the sense that the principles of physics and the engineering principles involved in technology transcend national boundaries. Scientists and engineers working within a discipline have means of communication that are not in any way limited by race, culture, religion, or environment. With the use of mathematics and translation,

language is almost no barrier. Many, if not most, scientists
strongly object to any national or cultural qualifying adjective
to science. Yet, though science and technology are universal,
there are also particularistic aspects to them. Science is a
body of ideas, some only distantly practical, if at all, while
others are directly useful in the human enterprise. Technology
is fundamentally problem—solving. The particularity of science
and technology derives from the fact that each country has its
own complement of problems and therefore must, in its science and
technology policy, draw and create from this common fund of
knowledge the most useful and relevant ideas and tools. What is
needed are national science and technology policies that
facilitate the organic growth of science and technology within a
culture. The universality and particularity of science and
technology can be the basis for finding unity in diversity and
diversity in unity. Modern science and technology have given us
that visual picture of our unity and diversity in the magnificent
photographs of the earth taken from the moon. Unity and
diversity in all aspects of the human endeavor are necessary if
we are going to share this globe and its heritage.

CHAPTER 3
Technology and Operational Principles of Transfer and Development

The purpose of historical inquiry on technology was to attempt to identify those operating principles that have governed technological evolution. This book refines, develops, and demonstrates these concepts. Consistency with these principles is necessary for successful technology transfer and economic development. They are fundamental, both in planning and in implementing technology projects and as a set of criteria for evaluating completed projects. There is technological science that has to be understood and utilized for development. Scholars in Europe have long distinguished between technology and technique (Sebestik 1983, 25-44). The present inquiry is more in the tradition of technology, although an attempt is made to incorporate certain basic techniques into the study. The next to last chapter takes those summary principles and combines them with the empirical and analytical results of current efforts in science and technology for development. The result is a set of guidelines or recipes for applying theory to the task of development.

The following are basic characteristics of technology:

1. Technology or human tool using is primarily an ideational process. It is the use of ideas to transform the material and nonmaterial world. The necessity for idea transfer implies that "science transfer must accompany technology transfer if the latter" is to take root in recipient countries (Salam 1983, 31). It is also the case that science transfer cannot occur without complementary technology transfer. Most scientific research today involves the use of sophisticated technology that must be transferred and maintained. Lack of adequate equipment or maintenance of existing equipment is a source of frustration for many third world scientists and one of the reasons that many choose to work in developed countries.

2. Technology is behavioral. The very existence of tools implies skills in both tool creation and tool use. Skills are forms of human behavior.

3. Technology becomes organizational and institutional. Humans live in groups, and much of their technology going back to

large-scale hunting requires organized group cooperation for effective implementation.

4. Technology as ideas (or knowledge) and as material artifacts is transmitted through culture. Though analytically separable, technology in use becomes part of the general belief system of those who use it. As such, the dynamic nature of technology can come into conflict with the restrictive institutional beliefs and practices.

5. Technology is cumulative and combinational. Once the process of technology is under way, it gains momentum from the ability to combine, recombine, and modify existing technology. In the same sense that biologists say all life comes from previous life and anthropologists say that all culture comes from previous cultures, we can state that all technology comes from previous technology. Though in terms of the archeological and industrial record, what appears to be combined are the material artifacts, the dynamics of the process come from combining the ideas of the artifacts.

6. Technology is an interactive process. Just as there are feedback loops between human evolution and tools, there are feedback loops with other social activities. If a distinction is made between science and technology (which we do not fully accept, although most writers on the subject do), then we can observe a continuous feedback loop. Scientific inquiry establishes principles that are applied in technology. Technology provides ideas and instrumentation that facilitate scientific advancement.

7. Technological change is an accelerating process. The more technology there is, the even greater are the possibilities for new combinations and advances, and for positive feedback loops between science (and knowledge in general) and technology. Goldschmidt (1967, 112-13) refers to "an exponential quality; that is, the rate at which growth takes place increases with each successive increment of advancement." He considers it "mathematically inevitable that the opportunity for new techniques increases geometrically as the number of basic ideas increases arithmetically." The historical record in the time scale figures that we have given certainly bear out the contention of exponential growth. It is a major contention of this book that some theories of technology and the projects derived from them neglect the potential of technology and thereby severely limit the potential gains from technology transfer.

8. Technology is a problem-solving process. Technology is technology in the context of its use. Its use in the wrong context does not deny its efficacy in appropriate circumstances. By definition, all technology, if it is truly technology, is appropriate to some problem-solving endeavor. The selection of technology depends upon cultural, environmental, and economic criteria that define a problem and the characteristics of its solutions.

9. Technological innovation has long-term consequences that, from the point of view of subsequent generations, are more important than the short-term problem solving that was

the original intention of the technological change. Both in the short and the long term, there are chance or fortuitous discoveries. Serendipity is the name given to the discovery that emerges by chance when one looks for one thing and finds another. As one great scientist said, "chance happens to the prepared mind" in that accidental discoveries only happen because someone had the intelligence to interpret the results as being useful in another context.

10. Because technology is combinational, it is not surprising that people working from the same technological and scientific base frequently create essentially the same invention (i.e., solve the same problem) at the same time. For the same reasons, simultaneous discovery is also a frequent phenomenon in scientific inquiry.

11. Technology, since it involves behavior, is likely to bring about cultural change. Similarly, culture is likely to force a modification of technology in the process of diffusion.

12. The cumulative body of knowledge, instruments, and human skills that we call technology is essentially a seamless web. The source of an instrument or innovation in one area of endeavor can be derived from what superficially appears to be a totally unrelated area. This point is illustrated in Chapter 6, the chapter on culture and technology.

13. Adaptation of technology to new environmental circumstances is in itself a form of invention. Socially and linguistically, we call large, apparently discrete technical changes inventions. A closer, detailed analysis of the inventive process reveals the fact that a series of smaller, less discrete changes preceded the more noticed, larger changes. A more in-depth study of the inventive process of printing would clearly indicate this. In most instances, adaptations are comparable to small, innovative changes in the inventive process.

14. Technologies that are separated from major areas of technological change, whether by geography, by culture, or by political isolation, tend to slow innovation, if not bring about stagnation. Linkage to other technologies--linkage that necessitates not only contact but also some compatibility in the level of technological achievement--is vital. This issue of linkage and continued technological development will be central to our analysis of appropriate technology.

15. Technology and science create resources. Resources are neither natural nor necessarily finite.

16. Historically, more often than not, successful technology transfer has involved borrowing technology at its then highest level.

17. Terms such as Western technology can have only a very restricted historical use. Modern technology can have a universal meaning as the set of knowledge and ideas, skills, tools, and machines that are the most efficient and effective at problem solving. Similarly, modernization, a term frequently in disrepute as being ethnocentric, can have a transcultural meaning as the societal and cultural ability to use modern technology.

18. Technology transfer is incomplete if particular

techniques are diffused out of context from the larger dynamics of the process. The presence or absence of a particular tool or machine is not evidence for technology transfer.

19. New technologies almost always allow people to do new things as well as doing old ones better. One of the mistakes of analyzing old and new technologies is that they are frequently compared only in terms of how well they perform established tasks.

20. In choosing between technologies, the borrower must recognize that the context is not the world as it was but the world as it is and as it is becoming, with most competitors (both economic and military) using the most efficient and most effective technologies.

21. Technology is an evolutionary process. Technologies are predicated upon the prior development of other technologies. This is the process in aggregate. For a particular economy or people, the process of technological evolution need not be replicated. Borrowing peoples may skip stages by borrowing technology at its highest level. As a universal process, technology is evolutionary. As a particular process, technology is revolutionary. The dynamic accelerated growth that characterizes universal technological change may be even more greatly accelerated when particular peoples jump stages of technological development. The very rapid growth of technology-borrowing countries, particularly the extraordinary, unprecedented post-World War II growth of countries such as the Republic of Korea, testifies to this process.

22. All technology is both universal and particular. It is universal in that it is based upon the principles of nature and is characteristic of all people. It is particular because as a problem-solving process each people has a unique set of environmental and culturally-defined problems to be solved. Because technology is universal, it is ironically and paradoxically also alien: for any particular people, the majority of their technology originated elsewhere. Simultaneously, technology as problem solving is entirely indigenous in that technology must be adapted to meet local conditions, and this process of adaptation makes it one's own. This apparent paradox of universality and particularity, alien and indigenous, when properly understood and explained, can facilitate technology transfer. Too often the alien character of technology for a recipient people is stressed, i.e., called Western technology, when it is also a universal technology that is the heritage of all human kind. Similarly, the "appropriate technology" movement stresses the particular character of technology and thereby ignores the enormous benefits of its universality.

23. Invention, as Veblen stated (1922, 314), is truly the mother of necessity. We adapt our lifestyle, population, and other aspects of our condition to new technologies. The agricultural systems of the world that create the means to support 4 billion people are necessary if there is not to be mass starvation. The technologies that allowed urban growth are now

necessary for their survival. Individuals may return to previous technologies, but rarely, if ever, does the group have that option. Going back to earlier technologies has been widely advocated as a development strategy, but no evidence has been offered as to its aggregate validity.

24. There are gains and losses to evolutionary and technological change. The human record shows that the gains have been far greater than the losses. The claims made for alternate technologies (and against modern technology) actually apply to modern technology. Nature is only a virtue in developed countries where technology has partially insulated the population from its worst ravagers, such as drought, famine, and plague. Modern technology is truly technology for the masses, giving people longer lives, more choices, and greater control over their destiny. The list of the benefits of modern technology is as long as the criticism against it.

25. Technologies coexist. New technologies are better for a range of problems and circumstances, but not for all problems and circumstances. Prior technologies frequently continue to experience improvements. New technologies often undergo further changes that may be more significant than the original "invention." Still, some technologies become obsolete and are a technological dead end.

26. Reliability can be substituted for redundancy. Modern technologies have greater reliability (i.e., redundancy) than do previous technologies. Evidence is overwhelming from data on famine and loss of life due to natural disaster. Most critics of modern technology cite breakdowns that are not substantially life threatening (New York City power failure) or theoretical possibilities.

27. Technological interdependence, or complementarity in complex technologies, means that some improvements in a technology require complementary advances in several areas. For example, fast-moving vehicles (cars, trains, planes) require braking systems. Toolmaking has always involved the property of materials. In modern technology, material properties (strength, purity, heat or corrosion resistance, etc.) have to be developed so that the proposed system can function.

28. The vital complementarities involve ideas and skills. Advances in the physics, engineering, and science of materials and in many other intellectual and applied inquiries are the foundation for the creation of the materials of modern civilization. As important are the advances in the skills necessary to operate the systems. Contrary to popular wisdom that we are losing our skills, modern technology has required that an ever-increasing proportion of our society be involved in intellectual and skilled endeavor.

29. Though there is an important sense in which modern weaponry and communications have linked the globe--tied us to one another's fate for good or for evil, and facilitated the growth of large centralized states, groupings of states, and international organizations--there is also a sense in which

modern technology is decentralizing. Very simply, it has given individuals and groups of individuals greater choice in the larger circumstances of their life and in the day-to-day range of choices available to them.

30. Last but not least, the free marketplace of ideas, democratic institutions of all kinds and free economic markets are all vital mechanisms in developing, transferring, and sustaining technology. The dynamic process of combining technologies to create new technologies or borrowing technology is greatly facilitated by freedom of thought and freedom of action. Jacob Bronowski, in his masterpiece, Science and Human Values (1959), argues that the basic principles of scientific inquiry as refined in the last few centuries are essentially the same principles of democracy--free and open inquiry are equally functional for science, for technology, and for democracy.

PART II

Technology and Human Endeavor

CHAPTER 4
Apocalypse Yesterday

"There is a question in the air, more sensed than seen, like the invisible approach of a distant storm, a question that I would hesitate to ask aloud did I not believe it existed unvoiced in the minds of many: 'Is there hope for man?'" Thus Robert Heilbroner (1974, 13), one of the most intelligent, sensitive, and perceptive economists in America today, begins his book, An Inquiry into the Human Prospect. Paul Ehrlich, probably the best-known prophet of doom in the 1960s and 1970s, is more emphatic than Heilbroner. He opens his book, The Population Bomb, with the assertion: "The battle to feed all of humanity is over. In the 1970s and 1980s hundreds of millions of people will starve to death in spite of any crash programs embarked upon now" (Ehrlich 1971, xi). Ehrlich and many of his illustrious contemporaries use the image of the Four Horsemen of the Apocalypse: war, famine, pestilence, and death. The very use of the apocalyptic metaphor by some authors for their prophecies implies that they have some concept of the long history of similar prophecies. These current prophecies differ, however, in that they are being made by scientists, using scientific evidence. It is reasonable, then, to use the scientific method of empirical testing on some of these authors' projections to give us at least a partial basis for assessing their recent pronouncements. As Garrett Hardin states (1978, 77), "To have any science at all we must generate falsifiable statements and have nothing to do with statements that are not falsifiable. Science is inherently vulnerable. It is proudly so." Hardin (1977, 1) also argues that "science does not admit invincible assertions into its sanctuary."

THE CATASTROPHISTS

Fortunately for our inquiry, some of the 1960s and early 1970s prophets of doom were highly specific as to the nature and time of the forthcoming disasters. Occasionally they hedged their bets by phrasing their forecasts as scenarios. Ehrlich, in one of his scenarios printed in 1969, specifies a series of catastrophes that were to be the lot of mankind during the 1970s: "The end of the oceans came late in the summer of 1979 and it came even more rapidly than the biologists had expected . . . By

43

September 1979, all important animal life in the ocean was
extinct . . . Earlier in the year, the bird population was
'decimated' (Ehrlich 1969a, 24-30). Along the way, 50 million a
year were dying of malnutrition; famine gripped many countries;
the green revolution was a failure as yields were falling;
200,000 a year were dying of pollution; the American Midwest was
turning into a desert; diseases of all kinds were on the
increase; and chaos was spreading. Amidst all these crises was a
baby boom in the United States. Ehrlich concedes that it is "a
pretty grim scenario, [but] unfortunately, we are a long way into
it." In his succeeding works, it is difficult to find any
indications of policy or other changes that were made to ward off
this catastrophe.

        In October 1979, one month after the extinction of oceanic
life was to occur, Paul Ehrlich accepted a $10,000 prize for the
best essay on the future. Three years later, the same prize was
won (now $30,000) for an essay that couldn't even get past deaths
correct. The authors had 500 million people die of famine in
1973-1974. This is quite an achievement since the deaths from
all causes in each of those two years was about 50 million. The
prizewinning essay was published (Freeman and Karen 1982, 184).

        Ehrlich published his prophecies widely and in some
surprising places, such as the Wall Street Journal and Reader's
Digest. He also called attention to other works of similar
orientation, such as Famine 1975! America's Decision: Who Will
Survive, by William and Paul Paddock (1967). The Paddocks were
as certain of the coming catastrophe as Ehrlich: "Nothing
can stop the locomotive in time. Collision is inevitable.
Catastrophe is foredoomed. . . . Now it is too late." There is
little potential for agricultural increase; hybrid wheats are
little used outside the U.S. (and presumably won't be); food
production is static. Famine--1975; catastrophe--1982. The four
horsemen are in their saddles and ready to ride. Though world
famine cannot be avoided, it can be "ameliorated" (Paddock and
Paddock 1976). Amelioration consists of saving ourselves and
some others, but not everyone. The Paddocks state boldly (pp.
207-9), "Herewith is a Proposal for the Use of American Food:
Triage." Triage is based upon the French wartime medical
practice of separating the wounded into three categories: (1)
those who would survive without immediate medical assistance,
(2) those who could survive but only with immediate medical
attention, and (3) those who could not survive no matter what was
done for them. Obviously, scarce medical personnel turned their
attention to the second group, then to the first, leaving the
third group to die. Triage as an international aid policy argues
that certain nations or peoples are doomed and that food
assistance merely delays and thereby worsens the inevitable
catastrophe. Further, trying to save the many could imperil the
future of the few. This view is essentially a variant of the
famed Lifeboat Ethic, i.e., if too many are allowed on board, it
sinks! There are, as Hardin (1977) says in another but similar
context, limits to altruism.

The year 1975 has come and gone, and so has 1982. It is difficult to find evidence for the Paddocks' predictions. Yet in their 1976 reprint (as <u>Time of Famines</u>), they confidently assert that "this volume demonstrates that it is possible to predict the course of at least some human events." They made no changes whatsoever in their text to prevent any perceptions that the data were massaged. In their postscript they accuse their critics of being blindly optimistic, guilty of "boosterism worthy of Babbitt," and simply not knowing what they are talking about. They also found the world's population to be growing at an accelerating rate, which it was not then and is not now. Unequivocally, they maintain that the Time of Famines is here.

Scientists such as Ehrlich and Hardin, and applied scientists with field experience such as the Paddocks (who are an economist and an agronomist), gave respectability to doomsday prophecies. Even more prestige was added when the Club of Rome asked a group of Massachusetts Institute of Technology (MIT) researchers to do a study of the human predicament using computers and the world dynamics approach of an MIT engineer, Jay Forrester. Among the results was the widely publicized book <u>The Limits to Growth</u> (Meadows et al. [1972] 1974). The tone of this volume is certainly not shrill, and their forecast for the apocalypse is on the order of 100 years, give or take a few decades, depending upon assumptions. Their general conclusion: "If the present growth trends in world population, industrialization, pollution, food production, and resource depletion continue unchanged, the limits to growth on this planet will be reached sometime within the next one hundred years" (p. 24).

When these limits are reached then the "most probable result will be a rather sudden and uncontrollable decline in both population and industrial capacity." <u>The Limits to Growth</u> study believes, however, that ecologically sound growth is sustainable far into the future provided that we change our ways. Of course, the sooner we change, the greater will be the chances of success.

Ehrlich, in defending the Paddocks' advocacy of triage, makes reference to Pascal's Wager (1968, 8; 1971, 179–80). Simply stated, Pascal, a French philosopher and mathematician, probability theorist and gambler, reasoned that belief in the Christian God entitled one to an afterlife of eternal bliss and that nonbelief meant damnation. If there was no God, both the believer and the nonbeliever were dead. The nonbeliever gained no benefit for being right but suffered a loss for being wrong, while the believer benefited from being right but suffered no loss as a consequence of being wrong. Consequently, there was everything to gain and nothing to lose by believing in God. Pascal also looked at the odds of eternal life against the short period of one's terrestrial life, so even if a price had to be paid, it was small in comparison to the gains. Similarly, Ehrlich argues that if the Paddocks (and Ehrlich himself) are right, we can expect disaster, and if we accept their rightness, then through triage and other actions we can ameliorate the forthcoming catastrophe.   Presumably, if they are wrong, there won't be a doomsday even if we practice triage.

There is a series of obvious fallacies in Ehrlich's reasoning.   Practicing triage is not costless, particularly if the doomsday forecasts are wrong.   If advocacy is successful in changing policy (and, after all, that is the object of advocacy), a very large number of people who might otherwise be helped would suffer privation and, possibly, death.   And what about those who create the practice of triage?   Keeping others out of the lifeboat isn't a pleasant thought and is fraught with moral and ethical implications, even when the lifeboat is full to capacity. If it is far from capacity, keeping others out is downright immoral.   Certainly, triage is not consistent with Ehrlich's (1971, 2) statement elsewhere that we "must all learn to identify with the plight of our less fortunate fellows on Spaceship Earth if we are to help both them and ourselves to survive."   The Pascal's Wager argument is equally applicable (or inapplicable) to any doomsday scenario and to the implied policies for its prevention.

Basically, Ehrlich (1969b, 140) favors foreign aid by such countries as the United States, with special emphasis on the "technology of birth control." He has spoken against "the export of death control," which one author, John R. Maddox (1972, 60), termed "paternalistically offensive."   Yet Ehrlich (1978b, 20) does find that Hardin in his book, The Limits to Altruism, arguing for inequality and privilege, "makes his case straight-forwardly and in some ways persuasively."   In that book Hardin is basically an opponent of aid to poor countries and makes a strong appeal for privilege both within poor countries and within the world community (1977, 46-69, 81-84).   Jay Forrester's work also has its political and ideological implications.   Forrester developed the systems dynamics approach that was the basis for the famed Meadows report commissioned by the Club of Rome on the limits to growth.   Harvey Simmons (1974, 199-201) quotes Forrester as saying that his work might "give the appearance of favoring upper income groups and industry at the expense of the underemployed."   Simmons finds Forrester to be impatient with democratic processes and in favor of policies that reduce health care and food supplies in order to save the World System from collapse.

POPULATION TRENDS AND FOOD SUPPLY

One of the ironies of the basic criticisms of doomsday forecasting is that the very policies criticized, the export of death control and food assistance, may be critical factors contributing to the solution or at least the amelioration of the population problem.   It is generally recognized that falling infant mortality rates will eventually lead to falling birth-rates.   Since a fall in birthrates begins later and initially falls more slowly, the first impact of decreased infant mortality and extended longevity is to increase the net reproduction rate (birthrates minus death rates), i.e., increase population.   In time the falling birthrate overtakes the death rate (or at least the rate of decline of birthrates is greater than the rate of decline of death rates), and what is called the demographic transition takes place as it has in the industrialized world.

In the 1960s it may have been naive optimism to believe it would take place in the third world countries. But from the mid-1970s on, there is increasing evidence that precisely this transition has been taking place. Even the Ehrlichs (1979, 70) now recognize that equity seems to be an essential factor in reducing birthrates: "When people are given access to the basics of life--adequate food, shelter, clothing, health care, education (particularly for women), and an opportunity to improve their well-being--they seem to be more willing to limit the size of their families."

The general world population picture is spotty, giving considerable grounds for hope and also for serious concern. A number of formerly poor countries have emerged as middle-income countries with life expectancies approaching those of industrial countries and with birthrates that have fallen dramatically in recent years. There is strong and increasing evidence of falling birthrates in most countries, even the poorest ones. Nevertheless, there is cause for concern because the evidence is weakest for the Indian subcontinent, where population control is needed badly, and for Africa, where in many areas there is no evidence of population slowdown. Recent world fertility surveys indicate that in most parts of the world women desire fewer children. Further, age at marriage is rising, reinforcing other trends for lowering birthrates. Many family planning programs are succeeding (Population Reports, 1979a, 1979b). Still, even if our most optimistic interpretations of the population data are correct, there still will remain a population problem. As every demographer (doomsday prophet or other) recognizes, even when fertility falls to net reproduction, the young age structure of a previously growing population can mean population growth for decades before it levels off. The current optimism about population growth is based upon declines in the rate of population increase.

The question is not whether we have problems of food supply and population, but what kind of problems and how bad they are. Many contend that the problem is not aggregate food production, but rather income distribution. D. S. Miller (1979, 323-24), in a review of Nutrition and the World Food Crisis, states: "Food requirements are about 2,000 kilocalories per day per man, woman, and child, and global agriculture produces something in excess of 4,500 kilocalories per head per day of crops suitable for human consumption. Much of this food is used inefficiently for feeding livestock to provide animal products for the rich, but even so there is about 2,500 kilocalories per head per day available for human consumption. So, please let us not hear about the 'World Food Crisis'." The author further argues that "all the evidence suggests that farmers can cope with the supply problem providing it is made worth their while."

There is in fact considerable evidence to support this last assertion. There was another study commissioned by the Club of Rome about the time of the Meadows report. This one was on the theoretical maximum food production possible in the world. The authors find that "taking into account the possibilities of irrigation and the limitations of crop production caused by local soils and climatic conditions, the absolute maximum production . . . is almost forty times the present cereal crop production" (Buringh 1975, 1). Calculating on the basis of 65% of the total of available land (65% being the percentage of the current land in production that is cereal crops), the potential output is 30 times the present production. The authors make a number of assumptions, such as use of the latest technologies, fertilizers, seeds and multiple cropping during all the growing days of the year. (In some cases their assumptions turn out to be too conservative. They assume that 80% of the arable land in Asia is in use; satellite data now indicate that in 1978 only 73% of land was being used for production [Rand McNally 1979, 49].) They recognize that in each specific area there are reasons why the absolute maximum cannot be reached. Even so, it does indicate a potential for agricultural development that is considerably beyond current production. Further, they assume no change in technology.

Some of the research efforts and possibilities recommended by the National Research Council, National Academy of Sciences (1977, 8-10), are simply astounding in their potentials. Among others are further possibilities for genetic manipulations of plants, development of nitrogen-fixing bacteria for cereal crops, improvement in photosynthesis, and production from currently unusable acidic soils. These are not pie-in-the-sky recommendations. Much of the research that the NRC called for has been under way for some time and shows signs of success.

RESOURCE TRENDS

On mineral resources, the same story prevails. Robert Solow (1976), in combating some of the fears of imminent resource exhaustion, gives U.S. government estimates for years of availability and the crustal abundance of some of the world's minerals (Table 4.1). General data of Solow on crustal abundance is supported by Vajk (1978, 64-65). Solow defines "known reserves" as those recoverable using current technology. Technology is continually creating new resources. Resources do not exist apart from technology. The raw materials of the universe become useful to humans and therefore resources as a result of technology change. Further, science and technology are, through processes of alloying, creating new materials. This allows for resource substitution in case of scarcity. As many writers have noted, long before one runs out of a resource, there are myriad possibilities for recycling and reprocessing waste.

One wonders, if we are being overwhelmed by population and are at the same time running out of resources, as the doomsayers insist, what has been the aggregate impact upon the world's

Table 4.1. Years available of world resources

Reserves	Known Reserves	Ultimately Recoverable Reserves[a]	Crustal Abundance
Coal	2,736	5,119	---
Copper	45	340	242×10^6
Iron	117	2,657	$1,815 \times 10^6$
Phosphorus	481	1,601	870×10^6
Molybdenum	65	630	422×10^6
Lead	10	162	82×10^6
Zinc	21	618	409×10^6
Sulphur	30	6,897	---
Uranium	50	8,455	$1,855 \times 10^6$
Aluminum	23	68,066	$38,500 \times 10^6$

[a] With current technology.

people? Despite all this travail, economic development has been unprecedented and has exceeded all forecasts.

In average per capita income the developing countries grew more rapidly between 1950 and 1975--3.4 percent a year--than either they or the developed countries had done in any comparable period in the past. They thereby exceeded both official goals and private expectations. That this growth was real and not simply statistical artifact may be seen in the progress that occurred simultaneously in various indexes of basic needs. Increases in life expectancy that required a century of economic development in the industrialized countries have been achieved in the developing world in two or three decades. Progress has been made in the world in the eradication of communicable diseases. And the proportion of adults in developing countries who are literate has increased substantially (Morawetz 1977, 67).

Data from The World Bank and other sources indicate that, though development has been slowed, it continued through the latter part of the 1970s (Bureau of Public Affairs 1979; World Development Report 1979). In the United States economic growth slowed during the 1970s, but it was not vastly lower than that of the 1960s. Despite the carcinogens in our food and the pollutants in the air, we added 2.3 years to our at-birth life expectancy, bringing it to 73.2 (Public Health Service 1979). The latest figures, for the year 1983, show a further advance to 74.7. This brings to a total of over 25 years U.S. gains in life expectancy in this century. With all the ways that we abuse modern technology, just think of the gains we can make (and have made) with its proper use.

Of all the resource questions, energy is the most complex and controversial. Energy shortages experienced in the United States in the 1970s were the result of political and pricing decisions and not of any resource shortage. Currently research is going on toward improved ways of using fossil fuels, such as coal; toward nuclear fusion; and toward solar, wind, ocean current, and geothermal sources of power. In the last few years with increased prices for oil and periodic uncertainty of availability, the oil-importing industrial countries have made significant improvements in energy conservation, particularly conservation in the form of increased efficiency in the use of energy. There is no lack of energy sources, merely a lack of the technologies to exploit them cheaply. The availability of cheap gas and oil has undoubtedly delayed work in this area, but there is reason to believe that eventually we will succeed here as we have done in other areas in the past. The belief that there is insufficient uranium for nuclear power in the United States or in the world just will not stand up to scrutiny (Deffeyes and MacGregor 1980, 66-76).

THE APOCALYPTIC VIEW

Many of the doomsday deadlines noted here have long passed, but this has not slowed the tempo of such prophecies. The Limits to Growth has the twenty-ninth day riddle. A "lily plant doubles in size each day. If the lily were allowed to grow unchecked, it would completely cover the pond in 30 days, choking off other forms of life in the water." On what day will the lily cover half the pond? The twenty-ninth. The twenty-ninth day riddle has become commonplace among those who are fearful of the future implications of current economic trends. Past prophecies may not have been precisely right, but we are warned that the pond is half full.

People in the United States and elsewhere are living longer at higher material standards of living and with more education than any previous generations have ever known. The question arises as to why the gloom about the human predicament and why the prophecies of catastrophe in the name of science? Though I do not pretend to have a complete answer, I will offer some comments that might contribute to a partial understanding. That we have a long history of apocalyptic predictions means that in some way this vision is part of our culture. Recently this apocalyptic vision has tended to be clearer and stronger on the coming catastrophe than on the new world that will arise from it. Religious visionaries in our society have both condemned modern science and tried to use it to buttress their beliefs. It is now understandable, then, that some segments of the public will respond to predictions of calamity by scientists. It is less understandable that scientists would make their sensational claims in terms that are alien to the ethics and mores of modern science.

In the predominantly protestant, Christian culture of the United States and the Christian culture of the European industrial countries, the work ethic has played an important ideological role in explaining and justifying inequality of wealth within a country and among nations. It has become obvious that we need not labor so hard to acquire our daily bread and that this may cause guilt pangs in terms of traditional beliefs.

If we are drawing our usufruct too easily, something must be wrong. The party is over. The price we paid is insufficient and so we will have to pay the full price some day soon. To use Ehrlich's baseball metaphor, "nature bats last."

T. C. Sinclair argues that there has long been a strain of religious thought in Western culture that opposed economic progress, particularly industrialization. Quoting R. W. Tawney, he notes that advancing wealth was viewed as bringing with it avarice, cupidity, and a weakening of traditional relationships. Sinclair (1974, 175) writes, "Much of the moral idealism which in earlier times found expression in various movements of social reform appears now, particularly in the USA, to seek an outlet in the environmentalist movement." Pollution has become a symbol of our moral degradation. We have sold our soul for affluence and the judgment day will soon be upon us.

ECONOMICS AND RESOURCE LIMITATIONS

In this author's judgment, there is considerable confusion on the issues of development and resource limitation, owing to a lack of understanding of the nature of technology and technological change. What the critics of modern technology fail to realize is that the main difference between current and earlier technological change and diffusion is one of scale. Modern technology is truly global in its impact. Modern technology is a problem-solving process, and problem solving generally creates other problems to be solved. Five centuries or more ago, before the development of flues, Europeans heated their homes with open fires (i.e., solved the problem of cold) and created massive amounts of pollution indoors. People throughout the world have (and many continue to do so) drawn their water from local sources that are used for animals and as the repository for various waste materials. They are using and drinking polluted water. Those lucky enough to have access to modern technology heat or cool their homes efficiently and drink (relatively) clean water. Few would argue that these problems are not solved better than they were previously, but in the process of solving them, we, the creators and users of the technology, have succeeded in creating air and water pollution that transcends the home and village and encompasses regions and continents, and is of near global proportions. To argue that modern technology creates pollution while earlier ones didn't is to deny the facts of real life experiences of most of the world's population. The difference is that our problems now transcend the home and village and must be solved at the global level; they can only be solved with modern technology. Twentieth-century populations with their problems and aspirations will not find solutions in nineteenth-century technologies.

The scale of modern technology is also a factor in technology transfer and the change it engenders. The rapidity and magnitude of the potential cultural changes involved in using modern technology are enough to overwhelm many cultures. But it also offers unparalleled opportunities not only for material improvement but also for cultural advancement. In sum, industrial technology dwarfs our previous technologies in its power to do good or evil. Those who would have us reject it are, to say the least, confused. The real choice lies in the opportunity to understand the nature of technological change and to use it intelligently to serve human purposes.

The economics model (meaning the basic body of theory and method used by most economists) argues that we have little to worry about from resource exhaustion. As a mineral or some other resource becomes more scarce, the price system acts as a mechanism for transmitting this information. As the price of a resource rises, we find ways of conserving it and we work to develop technology to use it more efficiently. One author has noted that as we move to greater resource scarcity, pollution becomes less of a problem because there will be less waste of resources (pollution) when the resources are scarce and their expense is great relative to the price of labor and capital (Rosenberg 1973, 116). Rising prices of one resource cause substitution of less scarce resources. It appears that in the biologist, ecologist, and Club of Rome models, feedback mechanisms tend to worsen problems. From Malthus onward, forecasts of doom almost invariably are based on a form of exponential growth that brings the end upon us suddenly and without warning. However, those who accept such forecasts will not be surprised by the end, in the way that those who accepted the faith wouldn't be surprised at either the apocalypse or the golden age that followed. In the economist model of the price system, most (if not all) of the feedback mechanisms are positive in the sense that they provide the signals for correction and redemption. Possibilities of serious general resource exhaustion are considered remote and virtually unlikely by most economists.

CO_2 AND THE ENVIRONMENT

There are other threats to human life that are not signalled to us through price. Groups convened by the U.S. National Academy of Sciences have concluded that the increasing CO_2 in the atmosphere is likely to lead to a warming of the climate and that chlorofluorocarbons (from aerosol sprays) are adversely affecting the ozone in the stratosphere and are thereby likely to increase skin cancer (Maugh 1979, 1167–68; Wade 1979, 912–13). Oddly, these are phenomena with a very large potential for damage, yet they receive far less publicity than other less well-founded prophecies. One presumably would be cheap to cure (there are alternate propellants for aerosols), while reducing pollution further (which we are doing) will be costly but necessary on a variety of grounds. Chlorofluorocarbons are outlawed in the United States but unfortunately they continue to be used else- where. One author, Professor Sylvan Wittwer, presenting a paper at the meetings of the American Association for the Advancement of Science (the publishers of Science magazine) argued that a warming of the globe might be beneficial to agriculture (Houston Chronicle 180). Then again it might not be. That is the nature of modern technology: its scale for good and destruction is so great that we do not want to take needless chances. The continuation of studies is the sensible step which we are pursuing.

The 1983 National Research Council study is careful in its conclusions on the impact of CO_2: "Our stance is conservative: we believe there is reason for caution, not panic." The report is full of uncertainties as to the totality of consequences of CO_2 and as to policy responses. As Schelling notes, the way we state the problem biases our conceptions of the solution. If it is a CO_2 or fossil fuel problem, then the obvious solution is to reduce fossil fuel consumption (Schelling 1983, 449). However, if we understand the problem in terms of climatic changes with differing beneficial and adverse consequences, then our range of possible best responses is significantly broadened.

Much of the concern about the warming effects of increased CO_2 in the atmosphere has derived from its possibly deleterious impact upon world agriculture. Interestingly, "the direct effects of more CO_2 in the air are beneficial: increase CO_2 around a prosperous leaf and it will assimilate more carbon and lose less water" (Waggoner 1983, 413). Carbon dioxide is a limiting factor in plant growth, so that increasing its availability improves photosynthesis. The indirect effects of climatic changes in the United States (warmer and drier) would likely be adverse, offsetting the gains from CO_2. The weather changes in other parts of the world would vary, with some of the colder areas, such as Canada and the Soviet Union, possibly experiencing longer more productive growing seasons. Summing all effects for the United States, Waggoner finds that the "wise forecast of yield, therefore, seems a combination of the incremental increases in production accomplished in the past generation as scientists and farmers adapt crops and husbandry to an environment that is slowly changing with the usual annual fluctuations around the trend."

Given current productivity and population trends, with a continued reduction in population growth, food production should be a diminishing component of world income. Consequently, with even a 10% or 20% increase in food production costs, in a world of "appreciably higher living standards, [the levels] that might have been achieved by 2083 in the absence of climate changes would be achieved instead in the late 2080s" (Schelling 1983, 475). The best estimates of the continuation of current trends of CO_2 in the atmosphere are hardly catastrophic for the next century. It is not at all clear that we can substantially change the trends without drastic changes in our economies. Given the great uncertainties in all our projections, the prudent course is to continue to study and monitor the situation and be prepared to take effective remedial action if our understanding of the trends warrants it. Further, we should be continuing our study for the opposite reason, namely that we might learn enough to where it might be feasible and beneficial to deliberately change the weather.

We cannot reject modern technology. There are no golden
ages in the past, and utopian visions of small-scale community
technologies in the future bear little relationship to reality.
To reap the benefits of modern science and technology, we must
act always carefully and intelligently. We must build in large
margins of environmental safety in the same way that we build
large margins of safety in our airplanes or our bridges. There
is evidence, in fact, that we are doing more of this than critics
concede. It is the affluence that technology brings that allows
us to set aside large areas for conservation. As Aldo Leopold
([1949] 1966, xvii) says in the famous conservationist work, A
Sand County Almanac, "Wild things . . . had little value until
mechanization assured us a good breakfast and until science
disclosed the drama of where they came from and how they live."
Possibly aprocryphal, but nevertheless illustrative, is the story
of the physicist testifying before the Joint Congressional
Committee set up after World War II to write legislation for
the control of atomic power. After hearing of its enormous
destructive powers, one of the senators asked if we couldn't
destroy all bombs and the knowledge necessary to make them. The
physicist answered, "Senator, the bomb is here to stay; the
question is, Is Man?" Most of us would love to rid the world of
destructive weapons and the knowledge to make them. We can't.
But we can control them. We can't rid ourselves of modern
science and technology because most people don't want to, and if
we were to turn back our technology very far the globe would
support far fewer of us. Nothing in this study is intended to
deny the serious nature of the world's population and resource
problems. I am in no way advocating complacency. However
difficult our problems may be, there are no empirical grounds
for apocalyptic visions of certain defeat. We could lose, but
the bulk of the evidence indicates that we probably won't. There
is a solid factual basis for cautious optimism. Success in
solving these problems is a function of the extent to which human
beings are willing to act with intelligence and not of inherent
environmental limitations.

Of all the metaphors for our modern predicament, the one of
Spaceship Earth seems most apt. It recognizes that we live
together in a system and must work together cooperatively so
that it functions for everyone's benefit. The Spaceship also
symbolizes human beings' ability to use science and technology
creatively to break barriers that were previously restraining.
Mankind is capable of soaring toward the stars. John Dewey
(1929, 3) sums up the arguments:

> Man who lives in a world of hazards is compelled to
> seek for security. He has sought to attain it in two ways.
> One of them began with an attempt to propitiate the powers
> which environ him and determine his destiny. . . . The other
> course is to invent arts and by their means turn the powers
> of nature to account; man constructs a fortress out of the
> very conditions and forces which threaten him. . . . This is

the method of changing the world through action, as the other is the method of changing the self in emotion and idea.

The choice is ours. Apocalypse did not happen yesterday. We need not give up on tomorrow or cower under visions of doom. Within broad limits, we are free to create a new tomorrow. To the question of whether man will continue, I answer with words from the 1950 Nobel Prize acceptance speech of William Faulkner who was generally more somber in his novelistic views: "I decline to accept the end of man. . . . I believe that man will not merely endure: he will prevail."

CHAPTER 5
Technology and the Use and Creation of Resources

The various catastrophic visions that saw the world exhausting energy resources, minerals, land, and breathable air were directed mainly towards the activities of the industrial countries. Yet, if global limits to growth had been reached, economic policies for both the industrial countries and the less developed would have had to take this into account. Much of the impetus for the appropriate technology movement of the 1970s came out of a belief that renewable resources had to be used because we were fast running out of the other kind. That the specific catastrophic visions have not been borne out does not negate the possibility that the forecasts on minerals may have been more accurate. This chapter addresses the question of the nature of mineral resources and the possibility of mineral availability for use in both developed and underdeveloped countries.

The first elements created were hydrogen and helium. With the formation of stars came the fusion processes (nucleo-synthesis) that created the elements up to iron. There the elements (and future minerals) remained trapped until the explosive death of some stars released them into the surrounding space. This process of explosive death is also responsible for the creation of the elements heavier than iron. Only a small percentage of stars end in this violent type of death, called a supernova. With the exception of hydrogen and helium, then, all the elements that constitute the earth, its minerals, and its living inhabitants are the ashes of long-dead stars.

The earth consists of a hot molten core of mainly iron and nickel and possibly sulfur, a solid but hot mantle containing most of the elements, and a crust 15–30 miles thick on land, considerably thinner under the ocean. In this crust, the 8 most common elements constitute about 98% of it by weight and even more of it by volume.

Of these 8, oxygen and silicon together comprise about 75% of the earth's crust by weight. Aluminum is a little over 8% and iron is 5%. The other 4, calcium, sodium, potassium, and magnesium, range from about 2% to just over 3.5%. These 8 elements alone, with carbon and some trace elements (nitrogen, oxygen, and hydrogen from the air, as well as oxygen and hydrogen for water) and energy from the sun have been the basis for life on earth. They have given life the land to roam on, plants the soils to grow in, and humans the solid materials to build with.

Only iron, of the first 8, has been a basic metal of civilization. Aluminum, because of the difficulty of freeing it from its compounds, is only a recent addition to the primary metals. It still remains energy intensive for processing, and most of the economically workable ores result from several million years of leaching out of other components from aluminum ores in tropical soils. Even though magnesium is one of the major minerals of the crust, the cheapest way to obtain it is from the sea. Most of the minerals of industrial civilization are relatively minor parts of the earth's crust. In fact, for most of the minerals for industrial use, the prevailing belief is that they were not in the original crust to any large extent. They are, from this perspective, intrusions into the crust from the mantle.

Though many now agree that there are no serious aggregate problems of resource shortage, either in the present or on the horizon as a distant threat, there is a serious and substantive disagreement as to the fundamental nature of the resource problem. These differing perspectives may seem abstrusely theoretical at times, but these differences are critical to the policy making process.

It is only fair to note that this author takes roughly the following view: ultimate resource exhaustion is unlikely. There are possibly a few resources with economic reserves that are in short supply, for which the physical supply is adequate for decades to come, but for which there are other possibilities for other restrictions.

Before exploring the details of minerals availability, it is necessary to develop further the functional theory of resources and the evidence for it. So much has been said about the limits-to-growth theory that many otherwise educated people seem unaware that there are different interpretations:

1. <u>Limits-to-Growth Theories</u>. The basic belief is that most resources, particularly minerals and nonsolar energy, are in some sense finite. The faster they are used, the sooner they will be exhausted. Exponential growth in resource use will result in sudden and catastrophic collapse. Increased efficiency in resource use, new technologies, recycling, and other forms of resource conservation will only postpone (generally for a decade or less) but not prevent collapse. The only long-term solution to finite resources is population control, zero economic growth, and a shift to renewable resources in all areas of human life. These and similar strategies for preventing resource exhaustion were considered to be fallacious by the Paley Commission (1952, vol. 1, 21) and were called "a hairshirt concept of conservation which makes it synonymous with hoarding."

2. <u>Critical Resources in Short Supply</u>. This position is consistent with (1) above or with (3) below. Certain key mineral or energy resources are in short supply and are likely to be

exhausted within a few decades. For many of their uses there are reasonable substitutes, but for some there are as yet no known substitutes. In our complex economic and technological structure it is difficult to assess the extent of the adverse effect of these resource losses, but they are likely to be significant. As noted, one can hold to the concept of critical resources as part of a larger theory of resource exhaustion. The belief here would be that some resources, such as iron and coal, might last a century or more, while others, such as lead, zinc, and mercury, might be exhausted in a decade or two. It is also possible to argue that, while some mineral or energy resources are near exhaustion, in the long run science and technology will develop substitute materials or new means of economically exploiting lower grades of this otherwise "exhausted" resource. However, there can be an interval between the economic exhaustion of an existing resource and the time in which new technology can create new economic resources. This interval can have a substantial adverse effect upon the ability to use existing technology to sustain life.

3. <u>The Functional or Cornucopian View of Resources</u>. One is unlikely to find many who would endorse the Panglossian view that there are absolutely no mineral or energy resource problems. The functional view argues that there are no such things as natural resources. Human intelligence creates resources. Nature, the physical environment, or whatever one chooses to call it, provides the building blocks and the limits within which human ingenuity operates to fashion a way of life. Resources at any one time are a function of science, technology, and perceived human needs. To what is probably a substantial majority of economists, the idea of resources being finite is essentially an uninteresting theory. It is like rediscovering the wheel, or, in this case, the second law of thermodynamics. It is not a question of whether resources are finite, but whether they are sufficiently limited so as to require policy changes because of imminent resource depletion.

One can believe that in the long run there are no significant resource problems and still recognize severe resource problems in the short run. As noted above, it is possible that resources generally might not be exhausted but there are, using current technology, some specific resources in short supply, or else they are so costly to obtain as to be uneconomic. In this case, mankind has a problem of conserving existing supplies and creating the technologies for new supplies or new material substitutes. The issue is one of creating new technology for a transition that allows for economic continuity. This contrasts with the limits-to-growth theories, which allow for no transitions because exhausted resources are truly exhausted and therefore require a fundamental redirection in economic and technological activity. There are no technological solutions. Attempting to solve resource problems with technology is illusory, a technical fix that won't work and in fact will make things worse. To the many limits-to-growth theorists, technology is the cause of the problem, not the solution.

It is also possible to argue that there are no serious global problems of resource exhaustion, even for specific reasons. However, some key minerals, including some of the most abundant, are distributed in high concentrations in some areas. Thus, there may be no global physical shortage of these economically exploitable resources; however, institutional factors, such as political conflict or simple economic blackmail, may disrupt the flow of resources and cripple other economies. This is true whether it be an adverse impact upon fertilizer output and food production in poor countries owing to a rising price of oil, or a potential crippling of industrial economies due to a lack of energy or essential minerals. Where military prowess is a function of science, technology, and industry, industrial disruption can have more serious consequences than merely economic loss, though that in itself can be serious.

It is, of course, possible to believe in both a future general resource exhaustion and/or a specific resource in short supply and still recognize that otherwise abundant resources are subject to restriction due to factors other than physical availability. However, as noted, the policy implications are likely to be quite different. To limits-to-growth theorists, disruptions in supply are often seen as a good thing in that they require us to begin redirecting economic activity, something nature would eventually force upon us. To them, the earlier the change, the better.

Using resources is not the same as using them up, though there seems to be much confusion on the subject. Equating resource use with ultimate exhaustion assumes that there is some fixed stock of resources (i.e., that resources are finite) and that any use diminishes the resource. This is ultimately a position without hope for the future, because any pattern of use (including recycling, which can never be 100%) will lead to eventual exhaustion. Doomsday can be delayed but not eliminated.

It probably makes more sense to speak of wise and unwise use of resources than of using them up. For it is the intelligent use of resources that creates more resources. It is the use of resources that gives rise to the incentive to find new deposits. And, of course, without the use of resources, there is neither the incentive nor the science and technology for the creation of new resources. "Resources are not, they become; they are not static but expand and contract in response to human wants and human actions" (Zimmermann 1951, 15).

There is considerable historical and scientific evidence for the cornucopian position on resources. It is true that people at times have run out of specific sources and that defense of one's own resources has been a causal factor for armies marching. Known sources of tin were exhausted in the Middle East several thousand years ago, threatening the Bronze Age civilizations of that time. Then, however, improved ways of working with iron

were found (Asimov 1979, 294–95). More recently, supplies of
whale oil were becoming increasingly scarce just as we were
learning to explore and drill for petroleum.

PREVAILING ECONOMISTS' VIEWS ON RESOURCE EXHAUSTION
      One of the most comprehensive works on the economics of
historical resource use in the United States is the 1963
study, Scarcity and Growth: The Economics of Natural Resource
Availability, by Harold J. Barnett and Chandler Morse. Barnett
and Morse (1963, 49) spell out the traditional view of resources:
"The belief seems to be that natural resources are scarce; that
the scarcity increases with the passage of time; and that
resource scarcity and its aggravation impair levels of living and
economic growth." To economists, increased scarcity would be
manifested in increased cost or price. Barnett and Morse put
this to the test and found it wanting. They found that the value
of extractive industries has fallen steadily relative to total
output from 1870 to 1957. Or, stated differently, the extractive
sector of the economy was using a continually decreasing proportion of the economy's labor and capital. Decreasing real costs
prevailed except for one resource industry: forestry. This is
ironic, since putative exhaustion of natural resources has led
some to argue for turning more to renewable resources, such as
wood.
      Later, works by William Nordhaus (1974, 22–26) came to
roughly the same conclusions. Looking at the price of minerals
relative to the price of labor for 11 minerals from 1900 to 1970
(data available only from 1940 and 1950 for 4 of them), he found
that the relative mineral and petroleum prices had fallen for all
of them, though in the 1960s the relative price of copper had
risen. Even with a decade-long rise in the price of copper, in
1970 the price of copper (relative to the hourly wage rate of
labor in manufacturing) was less than two-ninths of its 1900
price. Clearly, technological change had more than compensated
for a substantial decline in the copper content of the ore. This
fall in relative price occurred over a period in which the
average grade of copper ore had fallen to between one-fifth and
one-tenth (depending on whose data one used) of what it was in
1900. The extension of the Barnett and Morse data to 1980 shows
a continuation of the downward trend in real mineral prices
(Barnett et al. 1984, 325–26). If in the 1970s the relative
price of some minerals or other resources increased, it was not
necessarily the result of resource scarcities or difficulties of
extraction.
      Many of the projections of yearly supplies of resources are
a function of the patterns of investment, exploration, and, at
times, luck. In 1929 forecasts were made in the United States
indicating only a 10-year supply of lead. In 1952 the Paley
Commission was fearful about lead supplies keeping up with
demands (even with high prices), though their primary concern was
for domestic U.S. supplies. In 1972 The Limits to Growth study
found, using 1970 data, that there were from 21 years to 26 years

of lead reserves. By 1974 our reserves of this mineral lead (that we are supposedly exhausting) had risen to between 29 years and 47 years (Dearborn, 1980, 212). Current estimates of world lead reserves are generally still in excess of 20 years. In fact, in the last few decades the ratio of reserves to yearly use has increased for virtually all minerals and other resources. The most egregious and important exception is petroleum and natural gas. From the late 1940s to the late 1960s, reserves of iron ore increased 1221%; manganese, 27%; chromite, 675%; copper, 179%; and lead, 115% (Carman 1979, 78).

Diamonds are another instance where recent projections have been outstandingly pessimistic. Life expectancy projections in 1974, gave 8 (demand growing) and 9 (demand static at 1974 levels) years of reserves for industrial diamonds, which means that by now we have exhausted these reserves. However, world markets are currently glutted with all kinds of diamonds.

RESOURCE AVAILABILITY: THE CHANGING PICTURE

Even without improved technologies for resource exploitation, there are additions to resource reserves because of new finds. The process of resource discovery is also subject to scientific and technological advances (Guild 1976, 709-13). Advances in geology, such as plate tectonics, and in geography give us new understanding of how various ores are formed and sorted, and where they are likely to be found (The Economist 1980, 63-69). Remote sensing from satellites has added new dimensions to mineral discovery and mapping, though at times it is easier to be more enthusiastic about the potentials of remote sensing than is yet warranted by the data. Airborne and ground surveys, using a variety of technologies, aid in the search for new material sources.

Exponential economic growth does not necessarily mean a similar growth in the use of scarce mineral resources. For some minerals, resource intensity (quantity of resource used per unit of GNP, corrected for inflation) declines with economic growth. In advanced industrial countries the use of manganese for processing to eliminate iron sulfide ($Fe\ S_2$) becomes less intensive as electric furnaces replace open hearths for the basic oxidation process in steelmaking, though the use of manganese in creating alloys continues. The high temperatures of the electric furnaces remove more sulfur by oxidizing it. However, new uses for manganese could reverse these trends, particularly if seabed sources lower its relative cost. Also, with new technology, we learn to use resources more efficiently. For example, in 1900 it took 7 pounds of coal to produce 1 kilowatt hour of electricity, but by the 1960s it took only 0.9 of a pound (Rosenberg 1973, 116). Barbour et al. (1982, 34) write, "Efficient use of energy has been one of the most important conditions of material progress. . . . A diesel railroad engine did the same work as a coal-fired steam locomotive at one-sixth the energy consumption."

Also, transportation technology can make already useable resources available over a larger area. Writes Rosenberg (1980,

63), "Again as recently as the 1930s, natural gas was still regarded as an unavoidable and dangerous nuisance that needed to be safely disposed of . . . It required the perfection of the technique of producing high pressure pipelines to transform natural gas from a waste product."

Necessity can also force people to use resources more efficiently, as the shift to higher-mileage cars in the United States demonstrates. For example, in 1943 Germany was producing locomotives using one-tenth the amount of copper they were using in 1942. Though resource-short, Germany ended World War II with greater stocks of most raw materials than it had in 1939. This it achieved both by greater efficiency of use and by resource substitution (Mason [1978] 1980, 305).

The evolution of the computer over the past three decades or so further illustrates the resource-saving character of many technological advances. According to Isaac Asimov (1979, 355), one of the first computers, ENIAC, "contained 1,000 vacuum tubes, weighed 30 tons, took up 1,500 square feet of floor space, and used up as much energy as a locomotive. . . . [Now] a computer that consumes no more energy than a light bulb, that is small enough to be lifted easily, that can do far more, and is twenty times faster and thousands of times more reliable [is available] at almost any corner store." New advances in semiconductors promise even more spectacular resource savings in the next few years.

New technologies often allow for the production of a previously existing output by more efficient resource means. Optical fibers, using lasers for carrying messages, will be replacing copper in cables with silicon and oxygen, the two most abundant resources on the Earth. (Together they equal 75% of the crust by weight, over 90% by volume.) Even before they have been extensively used, optical fibers are being improved for efficiency by multiplexing messages. Of course, transmission by satellite has opened an entire world of communication and saved many resources that would have gone into laying cables. The rapidly multiplying uses of ceramics and the improvements in them constitute the creation of virtually limitless new resources (Bell 1984, 10-12).

New technologies give forth possibilities of exploiting new resources. Before the end of this decade the technology for obtaining rich sources of cobalt, maganese, and nickel will be operative, provided international political considerations make it feasible. For several decades we have been getting bromine and magnesium from sea water.

In some respects we will never "run out" of minerals. Except for those minuscule amounts that we may shoot into outer space, all materials used will exist in some form on earth. Some will be concentrated and available for reuse (i.e., recycling). Others will be sufficiently diffuse as to not make it economically worthwhile to concentrate them for reuse. The Limits to Growth estimated, for 1970, a 31-year supply of aluminum, assuming exponential growth (Meadows et al. [1972]

1974, 56-57). Aluminum by most calculations is approximately 8% of the earth's crust (Brobst 1978, 120). It is obvious from this and other illustrations that "exhausting" or "running out of" mineral resources are, at best, imprecise metaphors. In actual fact, what they mean is the nonavailability of a resource in a concentration or ore form that makes it economically usable with current technology. As we have noted, technological change alters resource availability. Most economists argue that price changes signal resource scarcity and allow people to make approximate adjustments.

Prices also alter resource availability. For example, the Paley Commission found that "at cost levels prevailing in 1951, our recoverable coal reserves amounted to some 30 million tons." At a price 50% higher, estimated resources were "twenty times greater" (Rosenberg 1973, 1). Thus, as always, it is important to note that reserve estimates are at current technology and current real prices. Increases in the real price of one mineral relative to others lead users to substitute other minerals and consumers to reduce demand for the final product as its price rises. Thus, price increases by increasing reserves and decreasing demand have an even greater impact on the number of years' resource supply (which is the ratio of reserves to yearly consumption). The price system as a signaling device for resource scarcity and individual adjustments to these signals is one of the reasons that most economists do not fear resource exhaustion.

There are some who argue that changing technology will eventually allow us to mine "average rock" for minerals. Brooks and Andrews ([1974] 1976), for instance, say "the literal notion of running out of mineral supplies is ridiculous. The entire planet is composed of minerals, and man can hardly mine himself out. . . . Therefore, it is simply not true, as is often remarked, that average rock will never be mined." As a matter of fact, right now there are instances where "we are already mining commodities as by-products whose average grade in the ore deposit is lower than that in the crust (titanium mined in beach sand is an example)." To utilize these ever-decreasing grades of ore down to common rock will require greater energy to extract the ore, to reclaim the land, since far greater volume will be processed for ore, and to counter the greater pollution from greater energy use itself.

In the future, not only will energy be one of the keys to mineral availability, but it will also provide an experience and model for understanding the minerals problem. Had this chapter been written in 1970 on energy sources, it would have been simple to point out the world dependency on petroleum from a few concentrated sources. One could have noted the long-term declining price of energy and the historic function of science and technology in replacing exhausted energy sources with new more productive ones. One could have noted the multiplicity of new, potentially inexhaustible sources, i.e., geothermal, tidal, solar, solar satellite, nuclear fission and fusion, etc. One

could have noted the possibilities for greater efficiencies and
substitutions, such as insulation for energy. One could have
noted possibilities of greater recovery of oil, new finds, or
obtaining oil from shale, tar, or heavy oil deposits. All of
this was true then and remains true today. Further, policies are
easier to frame in advance of a crisis than after it is upon us.

A look at some of the mineral needs as they changed from
1950 to 1975 and as they exist today indicates clearly why
minerals policies cannot be set in concrete but must be continually monitored to meet changing circumstances:

1.   In 1950, 95% of the world's production of columbium
(now generally called niobium, though the ore is still called
columbite) took place as a by-product of tin mining in Nigeria.
It was a high-temperature super-alloy for which "no satisfactory
substitute" was then known (Paley Commission 1952). By the mid-
1970's, all of Africa (including Nigeria) had fallen to third
place, behind Brazil and Canada, in supplying columbium to the
United States. Its uses have greatly expanded, but vanadium,
manganese, molybdenum, and tantalum are seen as substitutes
(Bureau of Mines 1975, 282-87).

2.   In 1950, the wide availability of silicon was commented
upon as well as the fact that it was little used. The use of
silicon of high purity for semiconductor studies was mentioned in
the Paley Report. Its importance today (and possibly even
greater if the future of solar power is developed) is of little
concern because of its widespread availability.

3.   In 1950, gallium was a scientific curiosity produced
and sold by the gram. The Paley Commission recognized that its
properties gave it great potential, but "new uses" were needed.
Today gallium (particularly in combination with arsenic) has so
many uses in electronics, where it is such a prime and superior
material, that it is difficult to think of lasers, diodes, or
semi-conductors without it.

4.   In 1950, titanium was just emerging as a significant
metal. It was "produced on a moderate industrial scale," though
it was already recognized that its use would expand greatly in
the coming decades. Today one rarely sees a list of strategic
minerals that does not include titanium and refer to its use with
aluminum for aircraft.

5.   In the Paley Report, tantalum is mentioned rarely and
only in conjunction with other metals. At the time the report
was issued, the U.S. government was already busily looking for
worldwide sources for it and columbium. For the past 30 years,
it has been a mineral in which there has been periodic shortage.
Since 1975 the U.S. government has been stockpiling it (Bureau of
Mines 1975, 1091-98).

6.   Vanadium is also treated lightly in the Paley Report
and referred to as "the least critical of all the alloying
elements used for steel making."

The point is simply stated: mineral needs and available
sources change, and minerals policies must change, too.

Science and technology are an important part of any minerals policy. In many respects, science and technology create the minerals problems that we call upon them to solve. However, we should note that far more problems are solved than created; the problems of science and technology emerge from their success, not from their failure. Historically, one strategem of avoiding resource scarcity was a movement toward the use of more different materials. Now we use most of them. A single telephone may contain "42 of the 92 elements provided by nature" (Chynoweth 1976, 123). Not only are we using a complex array of minerals and compounds, but we are using them in intricate systems for the very highly specific qualities that they possess. For some electrical equipment, Chynoweth (1976, 124) writes, there are involved "assemblies of intimately interacting materials, each material being carefully chosen and developed to optimize the overall performance of the component. In consequence, even a minor change in one of the materials used in a complex component might have a major effect on overall quality or performance of the component. In such circumstances, substitution of one material for another may be difficult or impossible. This is particularly true for equipment in which materials performance has been pushed to the limit." He adds, "The more a society depends on complex and sophisticated equipment, the more vulnerable it is to scarcities of certain key materials, even those used in very modest amounts."

The science and technology that through human history has created materials substitutes has also created complex technologies with very specific, difficult-to-substitute-for materials needs. This will continue to be true in the coming years. To some this vulnerability is an indicator of a larger necessity to return to simpler ways. For the vast majority this return to the past is seen as neither possible nor desirable. A minerals policy must be one that monitors and responds to technological changes and the material demands thereby generated. For the one thing that this chapter points toward is that mineral problems are a matter of being able to make critical transitions to new materials and/or material sources. Hindsight may give clearer vision than foresight, but intelligent foresight gives a far greater scope for effective action.

In a real sense, on minerals, the sky is the limit. Long before we have exhausted mineral opportunities on earth, humans will be mining the moon and asteroids for high-quality ores. Already the possibilities of manufacturing in space are becoming clear, as experiments can be conducted to create new alloys and materials under conditions of zero gravity. In time, manufacturing in space can become a self-sustaining endeavor using solar panels for energy and mineral resources from the moon and asteroids.

Some authors have spoken of three types of civilization: (1) the first, like ours, exploits the resources and energy of its planet; (2) the second exploits the resources and energy of its planetary system; and (3) the third utilizes the resources and energy of the galaxy. We would be considered a Type 1 civilization, possibly in transition to Type 2 (Shu 1982, 561). Mining minerals on the moon has a number of long-term advantages. Solar panels (constructed with lunar materials) could create the energy for mining and processing the minerals. Problems of pollution or waste materials would be trivial or nonexistent. High concentrations of minerals such as titanium and nickel can be found on the moon. By spectral analysis we can determine the mineral composition of asteroids. We then can pick out one (particularly one that crosses earth's orbit), put it into orbit around the earth or land it on the moon and, mine it to exhaustion. Though possibly a distant option, it is one that should not be neglected.

The ability to draw mineral resources from space probably will be of no immediate value in solving the problems of world poverty. It is probably not even of near-term benefit to the industrial countries. However, who would have guessed in 1960 at all of the benefits to less developed countries from using satellites orbiting the earth. The main immediate relevance of the extraterrestrial mineral potential is that it is one more argument against the theses that we are running out of resources. If we base development policies on resource exhaustions that are not likely to occur, then we are needlessly prolonging poverty. For the present, continued resource availability for the rich and poor nations alike, there has to be stable sustained international trade and cooperation.

Ironically, extraterrestrial sources of minerals are only of peripheral relevance in the near future, precisely because they are still so widely available from traditional sources. The same is true for minerals from Antarctica and for some minerals from sources such as the continental shelf. However, if the limits-to-growth theories were correct, then these would be vital and necessary mineral sources if development is to be sustained. Clearly, technological breakthroughs will eventually make these mineral sources economically competitive.

CHAPTER 6
Technology, Science, and Cultural Development

Distrust of new technologies is deeply rooted in Western culture. It could well be, as suggested in an earlier chapter, that some forms of resistance to new technology are endemic to the societal process of innovating technology. The very conceptual framework in which we have discussed technology, as ideas, as skills, and as behavior, involves a complex of institutional beliefs and practices, sentiments and symbols, to which people acquire an emotional attachment that makes them resistant to change. Though much of the criticism of modern technology emerges out of this western tradition, the movements seeking alternate technologies have found resonance and respectability among thinkers of vastly diverse cultural, institutional, and educational backgrounds. Criticism of modern technology in industrial countries is an affordable luxury for the affluent who rarely themselves forego the benefits of modern science and technology. To acquire the "natural" or "real," be it in construction with expensive stone or wood or in foods, eating only the rare or organically grown--these natural life-styles are expensive because the means for providing them are extremely limited, making it a way of life possible only for a very small portion of the world's population. The irony is that some of the practitioners of this lifestyle in the United States call it "voluntary simplicity" and are under the illusion that a variant of it is a prescription for the solution of world poverty (Frieden 1979, 181–83).

Unfortunately, the citizens of poorer countries do not have the surplus means to opt for technologies that are aesthetically pleasing to the affluent. Resistance to the most effective technologies, be it in food production, sanitation, or education, is a costly and potentially life-threatening endeavor. Yet there are a number of concerns that thoughtful leaders of third world nations have about technologies that they perceive as being essentially alien. These concerns do not always question the economic efficiency of the new technology. Nor are they necessarily addressed to the issues, discussed in previous chapters, of resource exhaustion or catastrophes from population growth. Rather, it is a fear of cultural transformation and value changes

67

throughout the lifeways of a people, from the arts to the family
to public and personal morality. Obviously, the fear of changes
wrought by new technology results from the belief that these
changes are likely to be adverse.

New technology, be it the product of internal evolutionary
development, a transfer from abroad, or some combination of
the two, carries with it implicit behavioral changes. In our
statement of the basic principles of technology, we argued the
relationship that tools imply skills and skills are a form of
behavior. Technology, as we have defined it, is also a system
of ideas. New ideas and new forms of behavior are often in
conflict with established systems of belief. Scientific and
technological changes can undercut systems of belief with
behavioral implications far beyond those necessary to carry on
the scientific and technological endeavor. Depending upon one's
perspective, this can be viewed as liberating or as a threat to
the very fabric of society. Thus, there are legitimate concerns
about technological change that cannot be dismissed as mere
technophobia.

A look at some historic instances of technological change
and their relationships to other dimensions of culture can give
us some insights into the nature of the process and the range of
possible consequence. That the examples are drawn from a
particular cultural and geographical heritage are a function of
the author's background and not of the unique importance of this
historical experience.

Interestingly, various episodes in the history of Western
technology and science that in retrospect are seen as having
promoted the expansion of knowledge and the arts were viewed with
suspicion at their introduction. From earliest forms of writing
to printing, to more modern technologies, there was always a fear
that something important would be lost as a result of the new
technology. Nevitt (1980, 218) writes,

> The bards, who taught the works of Homer by "harp" as
> the embodiment of the Greek culture of their day, opposed
> instruction in the alphabet and predicted that it would
> destroy aural memory. By the end of the fifth century
> B.C., however, this new technology, which could store and
> restore the memories of any tradition, already had gained
> acceptance and had converted bardic presentation into an art
> form. Greek phonetic literacy is the foundation of Western
> civilization; it created our archetypal forms of philosophy,
> science, art, and education.

For people to express phobias about technological change, by
implication there must be new technology in use causing changes
to which some are objecting. If it is being used, then the
technology has its advocates as well as its detractors. One of
the early Greek users of this technology of a phonetic alphabet
for writing was the great Athenian playwright Aeschylus. His
masterpiece, <u>Prometheus</u>, is a veritable paean in praise of the

science and technology that was so dramatically transforming Athens. The arts and crafts and sciences in agriculture, shipbuilding, construction, pottery, and sculpture were working a cultural revolution. According to Kahn (1970, 135), "the new energies released in this revolution and the problems generated by the revolution provided the impetus and subject matter for the Athenian drama of the 5th century."

Technology provided the surplus that allowed the society to support the arts and gave the population the leisure to experience and enjoy the artistic achievements. Science and technology were directly affecting the dramatic arts in other ways, such as the use of the mathematical principles of perspective scenery painting.

> In addition to the development of a theory of perspective, stage sets presuppose the availability of carpenters and painters equipped with tools and materials and specialized skills, the manufacture of devices like hinges and machinery for shifting flats. . . . These skills and devices (and affluence) were at hand as products of an Iron Age revolution, a technological revolution expanding and diversifying the social structure and threatening to undermine ancient institutions and traditional modes of thinking (Kahn 1970).

Nearly two thousand years later, there were similar objections to the use of printing by moveable type in Europe. It was argued that skills of penmanship would be lost. Since books were prohibitively expensive, learning continued to be by rote memorization. Relatively cheap printing and more widespread availability of reading matter would, it was thought, reduce the need for memorization and the strengthening of the mind that it implied (Priestly 1960, 4). It is a frequently stated theme that new technology will somehow atrophy a capability of ours, be it physical, mental, or moral.

We are in no sense arguing the neutrality of technological change. Nevertheless, technology can be understood as capability to do things, and technological progress would be enhanced capability. Most of the arts involve the use of some technology. Even those forms that do not initially involve technology, such as storytelling, can, as we have seen, be enhanced by new technologies and even lead to the creation of new art forms, such as drama. A technology such as writing, and then later, printing, gives us improved means to accumulate artistic experience, just as it establishes a basis for the cumulation of knowledge in other areas of the human enterprise. This cumulative artistic experience can be a stimulus to further creativity. As Jorge Luis Borges so poignantly expressed it, the book extends not only our memory, but also our imagination (Rybczynski 1983, 4).

A number of other authors have noted that the combinational possibilities of having access to many different authors is a stimulus to "creative acts" (Eisenstein 1983, 432-44).

Similarly, for scientific activity, Eisenstein observes that understanding the "great book of nature" is helped by having access to or "exchanging information by means of the 'little books of men'" (p. 186). She further argues that reducing the need for "slavish copying" contributed to the critical attitude that produced "a distrust of received opinion and a fresh look at evidence" (p. 2). This openness to inquiry is the very lifeblood of science. As noted elsewhere, more advanced technology substitutes reliability for redundancy. The technology of printing gave both redundancy and greater reliability. Errors were eventually weeded out as the quality of texts was improved. Having scholars in different places with identical copies of the same works created a common, usable intellectual heritage. Possibility of publication and distribution to a wider audience became a stimulus to creative efforts. With printed instructions and manuals, the basic knowledge of crafts and arts was eventually separated from the cults and mysteries that dominated oral traditions (p. 139). This was part of the transformation of the arts that united them with larger intellectual activities and gave them a respectability that they did not previously possess. Advances in technology are frequently seen as dehumanizing and depersonalizing. As Eisenstein shows, printing had exactly the opposite effect. Prior to printing, most book illustrators were known, if at all, only by their initials. The huge quantum increase in information meant that more of the creative people involved in book production became known and recognized for their distinctive styles. The same is true for creators in technology, science, and other arts and crafts whose preprinting predecessors are largely unknown. As Eisenstein (p. 134) states it:

> Every hand-copied book, it is sometimes said, "was a personal achievement." Actually, a great many hand-produced books were farmed out piecemeal to be copied and worked over by several hands. But even where a single hand runs from incipit to colophon and a full signature is given at the end, there is almost no trace of personality left by the presumably "personal achievement." Paradoxically, we must wait for impersonal type to replace handwriting and a standardized colophon to replace the individual signature, before singular experiences can be preserved for posterity and distinctive personalities can be permanently separated from the group or collective type.

TECHNOLOGY AND ART

The other arts of Western culture have derived new means from changing technology. In the fifteenth century, mathematics was the basis for perspective in painting. In the eighteenth and nineteenth centuries, Newton and Goethe contributed to our understanding of visual experience in their theories of optics (Opper 1973, 170; Bremmer and Prescott 1984, 38–42). In the nineteenth century, chemistry created new color possibilities, and putting paints in tubes facilitated painting outdoors. At the same time, a new art form, photography, was created. This was to be followed later with filmic arts and in our time with the many forms of imaging used in science but having aesthetic merit. One could discourse freely about almost any of the arts and establish a positive relationship between science and technology and the arts. Who can even begin to understand the history of architecture and architectural achievement without comprehending the history of construction technologies and materials?

Some forms of construction, such as bridges, are simultaneous achievements in engineering and art. It has often been said that it is almost impossible to conceive of an ugly suspension bridge. The industrial technology that created vast quantities of iron and, later, steel allowed for the creation of bridges that were beautiful, safe, efficient, and resource-saving. Billington (1983, 30) writes, "Early cast iron was about five times as strong as wood and hence required one-fifth the amount of material to carry the same load. The drastic reduction in quantity of material allowed the design to let more water flow past the bridge during a flood."

There is no evidence that modern technology has diminished either the creation of art or its public appreciation in highly technological societies. In a country such as the United States all the statistical data, from museum attendance, to independent theater, to spectacular growth in ballet productions and attendance, point to a growing enthusiasm for the arts. If critics of modern science and technology wish to argue a decline in the quality of contemporary arts, then they have a more subtle and difficult case to make. There are some elitists who argue that "mass culture" is a contradiction in terms; they attack the modern experience for the very success it has achieved in opening access to art forms that have been virtually the esoteric domain of a select few.

For the performing arts, technological developments occurring after their creation can substantially expand the potential experience. Until the metal-braced (and later framed) piano came along, the performance of some musical compositions caused vibrations in the piano wires that could literally destroy it. The amount of tensile stress on the strings of modern pianos is on the order of 30 tons or more. Artificial fibers gave greater artistic control over the sound obtainable from stringed instruments. Though it may be true that the best violins today are those made in the late 17th and early 18th century around Cremona, Italy, those still played are strung with modern fibers. Further, says Hutchins (1981, 171-86), "every violin made before 1800 in general use today has been considerably altered because early in the nineteenth century there was a demand for more power from violins."

In retrospect, one of the attributes of great artists is that they have used the available technology to its limits and even strived to go beyond them. Harold Schonberg (1978, 15, 32), the then music critic of the New York Times, argued that

"Beethoven asked for impossible things from the players of his day. Should we in 1978 adhere slavishly to what has come to us in the printed score, or should we try to achieve the composer's intent, filling in the gaps that Beethoven's day could not achieve? . . . Often scales are cut short because of limitations of the early nineteenth century piano. Do we play as written or do we extend the registers? Most pianists today extend."

Great artists not only use new technologies to create new aesthetic experiences, they create new ways or concepts for achieving artistic effects. Musical instruments are clearly technological inventions, as is the way of putting them together, called an orchestra, as is also a type of composition for them, called a symphony. The development of the arts partakes of the same qualities of cumulative expansion of capability, as does the larger technological process. The cumulative nature of the process is not negated by the fact that artists will forsake a technique such as perspective in painting and find new possibilities in an earlier technique. Scientists sometimes resurrect a neglected theory, and sometimes new uses are found for old tools. As this volume's principles of technology state, technologies coexist. So do artistic styles. Nevertheless, the process remains cumulative, giving ever-increasing choices to the artist and to the audience.

The cumulative nature of the process does not necessarily confer greater achievement by the current generation. Most of us could not create a simple stone tool. The art of flint knapping is lost to all but a few archeologists. Similarly, many of us, even with the best of modern tools, could not match craftsmen of centuries ago in making furniture or other useful items. Within our modern society we do have those with the skills, craft, and improved technology to match and surpass their earlier counterparts. In the arts, as in other endeavors, there are always those rare towering geniuses that seem to cast a shadow over their own times and a good portion of the future. Clearly, geniuses in science, technology, or the arts are products of their time; what they do would be different were they born in a different time or place, yet they always seem to go beyond it.

What is proposed here is a weak technological determinism. An analogy is in order. A later chapter argues that an increase in per capita food supply has led in aggregate to more people being better fed. Nevertheless, within this aggregate there are instances or countries where the new miracle crops are leading to greater inequality and even a worsening of the condition of the poor. The weak determinism argues that new food-producing capability is most likely to lead to more food for more people. This has happened. Where it has not happened, the cause has been policy decisions, not an inherent negative factor in the technology. Similarly, greater technological capability in artistic creation is very likely to lead to expanded and enhanced artistic expression in a group of people. Further, if this enhanced capability is not put to creative use, it would not be the fault of the technology. Basically, we have a faith in the creative powers of peoples and cultures to use new technologies to carry forward and express their traditional themes and styles.

The theory of technological change and the arts (and economy and culture in general) has been defined as one of a weak determinism. It is an attempt to avoid the pitfalls of a theory of progress that admits of no exceptions. The twentieth century is marked by a sufficient number of episodes of the use of modern technology for mass extermination to dampen one's unqualified enthusiasm for the modern way of life. The hatreds and irrational forces that lead to such vile conduct are of long standing in the human endeavor and are in no way created by modern science and technology. Quite the contrary, the advancement of knowledge tends to sweep away prejudices. Evils such as racism are falsified by the light of scientific inquiry. That these pernicious doctrines are still with us and that they have at times gained institutional control over technological means of destruction is the major horror of our time. Calling for a moratorium on science and technology, as some have done, is both naive and counterproductive. It is naive because the political means to do so on a global basis do not exist. The best we can achieve is weapons control by treaty, not by technology control. A so-called moratorium on science and technology would in effect restrain the very means of problem-solving that offer the best hope of survival. As Madame Curie put it, "Nothing in life is to be feared—it is to be understood" (Hellman 1976, 135). In the seamless web of human inquiry and curiosity, where could we have permissable research where we could be certain that there were no serendipitously destructive possibilities? As the existentialist philosophers have told us, we can choose many things, but we cannot choose not to choose, because that in itself is a choice. Turning one's back on greater freedom to choose is one of the worst choices a people can make.

THE PROBLEM OF DIFFUSION

Thus far we have looked at the development of technology and the arts, primarily as it took place in the context of what is called Western Civilization. Of course, this process involved diffusion among the different cultural areas within this larger grouping and diffusions into and out of this region. The "Western" heritage of technologies that facilitated the evolution of western art forms was as much a product of the heritage of mankind as was technology generally as we have discussed it. Just as industrialization became known as "Western technology," many of the cultural artifacts and attributes of these industrial countries became part of a process called Westernization when diffused to third world countries. It has also been called modernization, a term in considerable disrepute in many quarters. There are also a variety of pejoratives, from cultural imperialism to colonization.

Many liberal, enlightened, and sensitive individuals in industrial countries have expressed concern about the dilution of

other cultures or the loss of cultural identity.   Still others
bemoan the loss of diversity in the world.   Much of what is
written today is almost exclusively on the negatives of cultural
diffusion.   It is important, therefore, to make some distinctions
on diffusion and the range of possible outcomes.

The literature of anthropology is replete with case studies
of small groups, generally very much less developed technologically than those with whom they have come into contact, who
have suffered severe cultural and social disruptions because of
the diffusion of even simple technologies.   Comparable tragic
transformations have occurred when migration has led to a
people's virtual inundation by others who have come to occupy
and dominate the land.   Such incursions and the accompanying
potential potential for obliteration are continuing today.   One
cannot rationalize this process by noting that throughout history
many identifiable groups have been submerged in larger groups or
have otherwise ceased to exist.   After all, in history some
people have wantonly slaughtered others.   No reasonable person
would use this as a basis to condone a contemporary massacre.
The extinction of unique cultural and linguistic identities in
the modern world is clearly to be deplored.   Some Indian groups
in North and Central America and in the Amazon, and groups in
places as widely scattered as Africa, the Phillipines, and New
Guinea, are threatened.   In Africa, a switch in language and
with it, cultural identity, appears in many instances to be a
voluntary process.

Well-intentioned outsiders can retard the cultural evolution
of a people by trying to preserve their culture.   Such attempts
at preservation are frequently made in a region by people with
"no prior contact with its cultural traditions" (Whisnant 1983,
61).   Whisnant specifically refers to the Appalachian region of
the United States, but his observations are widely applicable
throughout the globe and to the general concerns of this chapter.
In a passage dripping with sarcasm, he refers to the "cultural
parent-teacher-saviors" who would preserve the "sweet cultural
children" from the harsh modern world [p. 247].   This clearly
does not apply to many who today are working to preserve and
protect threatened peoples and cultures around the world, but it
may in fact be the unintended consequence of protective policies
that derive from misguided assumptions.   To Whisnant, the culture
that is perceived by the intervenor is rarely congruent with
the culture that is actually there.   He adds, "That cultural
intervenors may be on the whole decent, well-meaning people, even
altruistic people, does not (indeed must not) excuse them from
historical judgment" [p. 263].

Many of the "ancient traditions" that outsiders attempt to
preserve turn out on close inspection to be of rather recent
vintage (Hobsbaum and Ranger 1983).   Many manufactured traditions
were in their own time ways of attempting to change traditions
no different than the hybrid forms that emerge today when people
adapt to external influences.   There is always the potential
problem of paternalism when outsiders become involved in the

preservation of cultural traditions. Whisnant gives many examples for Appalachia of external cultural items that are now
presented as indigenous and refers to what he calls "traditional
chic," the application of traditional styles to the making of new
products rarely if ever seen in mountain homes, such as cloth
napkins, table runners, and placemats [Whisnant, 67, 260].

Maintaining ties to traditional culture can be costly
(economic or otherwise) to the local inhabitants. Whisnant
argues that except in extreme circumstances (war, massive
political repression, forced relocation), "people appear not
to conceive of their own culture and traditions as being nearly
as fragile" as others assume them to be (p. 261). One could
go on quoting Whisnant's fine book, which through a cultural
change in the U.S. is highly applicable to problems of cultural
change throughout the world today. Protecting people from
change neither preserves diversity (diversity for the outsider,
not for the local inhabitants) nor the intended recipient
culture. Outsiders can play many constructive roles, from
helping to rehabilitate decaying traditional structures to
rediscovering prior cultural art forms. The most important
function of external agents be they state or international groups
or agencies, is to allow the population the room to maneuver: to
adapt, to change, to preserve, and to create whatever mix of
traditions, styles, and economic activities are necessary to its
survival.

It is naive in the extreme to believe that appropriate
technologies are somehow the deus ex machina device to preserve
the integrity of small cultures. When the threat to these people
is the taking of their land, appropriate technology will not be
much of a defense. Nor will the conversion of the nation state
to an appropriate technology strategy miraculously end the
rapacity of its citizens against minority cultures. Nor will
it solve the population growth and land hunger problems that
frequently form the core of the economic forces leading to
colonization and population displacement and/or destruction. In
the tropics, the forces that are threatening population groups
are also destroying the rain forest habitat and causing the
extinction of plants and animals. Later chapters argue that the
best way to preserve the rain forest is with the use of science
and technology to increase agricultural output in the already
settled areas. The same tactics, if effective, provide at least
a temporary lessening of pressures. Beyond that, we clearly
admit to not having a solution to a very pressing contemporary
problem. Without being cynical, we argue that the small-is-
beautiful proponents have more the illusion of a solution than
the reality of one.

Where the loss of identity is the result of continuing
contact, keeping one group technologically backward is no recipe
for cultural integrity. Technological backwardness, whether
imposed or the result of voluntary policy decisions, can be
falsely perceived as an inherent and necessary attribute of a
culture. Historically, few things have led so readily to younger

people's abandoning the lifeways of their progenitors than the belief that these ways are out of keeping with the times and are a prescription for the perpetuation of poverty and powerlessness. Further, there is no way to insulate small groups from the influences of the larger world and the changes that inevitably follow. Even if one could, it may represent the preservation of diversity to the traveling social scientist, but to the members, life in a cultural museum may not be the most exciting prospect.

For the larger groups of the third world, which constitute the vast majority of the population, the fears of artistic degradation, anomie, and cultural destruction are, to say the least, greatly exaggerated. It is not the purpose of the expatriate social scientist or author to impose cultural change upon others in the world. However, we can talk about alternate development and technological policies and their consequence. If causal relationships are posited that we believe to be demonstrably false, then we have an obligation to demonstrate the falseness. Arguing causality or false causality will obviously influence policy choices to the extent one's argument is accepted. Though by the tenor of our argument it is clear we are advocating a technology, we are in no sense trying to impose it.

The interaction between Western countries and the rest of the world has been more fruitful for the arts than is often recognized. Many of the technologies discussed earlier in the chapter as having so greatly stimulated the arts had their origins outside of Europe. Continuing since then, there has been positive interaction. The influence of Japanese ukiyo-e prints on painters such as Van Gogh is evident, as is the early twentieth century impact of African art on the Cubists (O'Neil 1984; Rubin 1984). In more recent times, the art of the Indians of the Northwest Coast of the United States and Canada has helped to shape the arts of the Surrealists. Despite the frequent destructive impact upon local crafts, alien technology from Western countries sometimes brought improved tools for sculpture for the Northwest Coast Indians, or cheaper iron for the West African metal workers, and created new markets for artists elsewhere.

Tourism and other forms of widening the market are generally seen as being destructive of traditional crafts and values. However, there are instances where the opening of new marketing opportunities has encouraged the growth of indigenous artistic traditions. The technology of transportation has expanded the market for local crafts by bringing people to the crafts and facilitating the sale of crafts around the globe. Even though new technologies may not always prevail in this fruitful interaction between local traditions and international markets, it occurs frequently enough to make it an option worth considering. A recent article on the Otavalo Indians of Ecuador illustrates these possibilities.

Elsewhere in Latin America, most Indian groups are fighting a losing battle to preserve their cultures amid

poverty, discrimination and exploitation. In contrast, the
Otavalos have defied the stereotype, discovering that
economic success has served to reinforce their Indian
identity . . .

Commercial weaving not only has raised the living
standards of many Otavalos, but also has transformed the
relationships that Indians in Latin America normally have
with whites and people of mixed blood. The Otavalos are
proud and self-confident, and Ecuadorean society treats
them--though not the country's other two million Indians--
with special respect (Riding 1984).

Riding goes on to describe the "sweat shops" in which the
Otavalo worked for nearly 3 centuries, and their earlier virtual
enslavement by the Spanish conquest. Economic success has not
eroded Indian customs. "Otavalos who have moved to Quito or
who travel abroad on business always wear traditional dress
and preserve the Quechua language." Some of the Otavalo weaving
uses acrylic rather than wool. In many parts of the world,
"traditional crafts" use artificial fibers, aniline dyes, and
other products of industrial civilization. Many of the artistic
effects, such as particular colors or brightness, could not be
achieved without them. Thus, though these crafts are not traditional in the sense of precontact with industrial civilization,
they nevertheless are distinctive and they reflect the culture
and its aesthetic values.

In other parts of Ecuador and throughout Latin America,
Ariel Dorfman (1984, 3) found that literacy programs help the
poor and ethnic minorities preserve and enhance their traditional
culture. This is so even in cases where literacy training
originally was in a politicaly dominant language such as Spanish,
rather than in the native Indian language or dialect. In such
instances, technology did not ossify but it did preserve and
enhance cultural values.

New art forms, such as the novel, or new technology, such as
film, have given artists new means to express traditional themes
of their culture. The West African griot has had his work
extended to a larger audience, preserved in the writings of a
Camara Laye, just as printing earlier extended the words of the
bards of ancient Greece. In music, millions of people have seen
the film of a concert where Ravi Shankar played rock on a sitar.
There have been musical performances involving the sarod (played
by Shankar's brother-in-law, Ali Akbar Khan) and a violin (Palmer
1979). Our modern transportation is bringing performing artists
together across cultures and traditions, and concerts, tapes,
records, and radio are carrying discriminating, potentially
fruitful influences. Would those who bemoan the adverse
influence of exotic technology argue that there is a shortage of
creative artists in the third world? The diffusion of technology
works like international development; it expands the capability
of artists to express traditional themes.

A few years ago, a coworker in an Agency for International
Development (AID) program distributed an article entitled,

"Chinese Kosher Pizza." This referred to an item he saw in a
restaurant advertisement. His plea was to preserve diversity in
the world. It is interesting that he chose foodstuffs for his
illustrative metaphor. Clearly anyone in the United States (or
most any place else) who pretends to have a sense of good taste
is offended by the idea of Chinese kosher pizza. However, is the
idea as patently bizarre as the author assumes? After all, if we
take wheat domesticated in the Middle East, form it into noodles,
an idea developed in China or Southeast Asia, then add a sauce
with many ingredients, including tomatoes domesticated in the New
World, we have Chinese – Middle Eastern Amer-Indian noodles. In
Italy pasta fixed in this and other ways is called spaghetti and
is considered a national dish. In India there is an edifice
built by Hindu craftsmen under Moguls, using Persian, Turkish,
and maybe a Venetian architect. That conglomeration produced
the Taj Mahal. Last summer in Pakistan I was asked if I had
ever eaten okra. This West African domesticate is in fact
a major food of Blacks and Cajuns in my area of Texas and
Louisiana. In all 3 areas, Texas/Louisiana, Africa, and
Pakistan, it is prepared and tastes differently. It adds to
the diversity of diet and is considered part of the cuisine.
Diffusion clearly added to diversity within the cultures, did not
diminish it between cultures, and certainly did not diminish
their distinctiveness.

Great national cuisines are frequently the result of an
indigenous synthesis of diverse traditions. The cuisine of
Indonesia is said to derive from great external influences:
Indian, Chinese, Arabian, and Dutch Colonial (Marks 1981, 4).
The same author (quoted in Claiborne 1984) found the cuisine of
the Jewish community of Calcutta to be a "mixture of Baghdad
cooking married to the flavors of India" and that of Tunisia to
include French, Ottoman, and Roman influences.

Technological diffusion will change cultures and their arts
and crafts, just as technological change within cultures has
brought about enormous changes. The argument for technology
transfer is not the naive one that the economy will grow and
poverty will be reduced, but that everything else important will
remain the same. It is not a question of preserving cultural
practices but rather of giving them the freedom and opportunity
to grow on their own terms, meeting the needs of the populace.
The Industrial Revolution did not make its participants less
English or French, or whatever their cultural heritage was. Nor
did it diminish their artistic achievement (if our argument above
is correct). Quite likely, an English or French person of
several centuries ago might not consider the modern version of
the culture to be "true," but what is important, the contemporary
members do. The task before people of any culture is not to keep
it static, but to make sure that in the process of change the
culture remains coherent, distinctive, and satisfying to its
members.

When we speak of preserving the distinctiveness of a
culture, we are speaking of distinctive styles and subtle tones

and shadings of beliefs and practices. The substantive content of most of what we call our own is an amalgram of items that had their origins throughout the globe. There are no pure cultures any more than there are pure races or pure languages. These defining styles and themes can as readily be perpetuated (if not more so) with new technology than with obsolete technologies.

In a famous passage, the anthropologist Ralph Linton (1963, 326-27) has a "solid American citizen" awakening "in a bed built on a pattern which originated in the Near East but was modified in Northern Europe before it was transmitted to America." Linton goes on to describe the bed covers (origins either Near East or China), his slippers (North American Indian), and other garments (India) and his morning toilet (soap from Gaul, shaving from Egypt). Linton has our American dress, buy newspapers, and have breakfast. Each activity involves items with similarly diverse backgrounds, as do those of the rest of the day. Linton closes with the observation that "as he absorbs the accounts of foreign troubles, he will, if he is a good conservative citizen, thank a Hebrew deity in an Indo-European language that he is 100% American." For our American citizen or for any other citizen it is not where the items originated but whether one has command over them and makes them one's own. Like the paradox of technology argued earlier, all items are simultaneously alien and indigenous. Alien, since most of one's cultural baggage originated elsewhere; indigenous, because it is one's own. It is only when diffusion is recent and massive that there is a substantial concern over alien threats to the culture.

Before leaving the topic of preservation of culture, a brief reference to science, technology, and preservation of another kind is in order. It is frequently written that pollution from industry is destroying some of the great monumental architectural achievements, such as the Parthenon or the Taj Mahal. This is tragic and is certainly a mark against the way we use our industrial technology. The other side of the coin is that modern science not only preserves and protects, it also rediscovers and restores. Many of the great monuments not only fell into disuse, they were ravaged and used for building materials by local inhabitants. Archeology has recreated a past that was never recorded, and other sciences have traced our biological heritage back to earliest times. Great monuments have been rebuilt and artifacts have been recovered from the ground of other places and sequestered in museums. Some of these items were looted from their original habitat, and the move to return them to home countries is to be encouraged. The operation of an art museum involves the very careful use of modern science and technology in cleaning paintings, X-raying them, restoring them, and controlling the temperature and humidity of their environment (Schneider 1980, The Economist 1982, and Robbins 1984). Records, tapes, and films, etc., allow us to preserve the work of a performing artist after death. Books, prints, photographs, museums, travels, etc., give a contemporary artist a greater sense of the achievement of his or other cultures than artists have ever known

before.   In every area of the arts and humanities, science and
technology are demonstrating their capacity to restore, preserve,
and enhance.

TECHNOLOGY AND DECENTRALIZATION
     One of the concerns of those involved in development is that
the adoption of more sophisticated technologies will lead to
greater   societal   centralization.   Conversely,   one   of   the
arguments for appropriate or alternative technology is that it
is decentralizing.   Other chapters argue that sophisticated
technologies are decentralizing on the consumption side in their
capability to deliver benefits to those most in need.   This can
and has happened.   Further, technologies that are centralized in
some respects, such as modern generation of electricity, are
decentralizing in the way they deliver power, both for production
and for consumption.   Similarly, in terms of political structure,
large centralized democratic states decentralize decision making
to the individual, who has the freedom of choice to travel over a
wide area without a passport, customs, or change of currency.
     This thesis is simple--that the trend of life in democratic
societies using the most advanced technologies is toward more
and more effective decentralization of decision making.   The
technology of printing was clearly decentralizing.   The wider the
distribution of books, the more people that have access to them.
The technology of abundance has a strong decentralizing tendency
to it even though size, cost, or other factors may partially
counteract it.   This decentralization is taking place throughout
the lifeways of people in technologically advanced countries.
The range of choices of artistic experience to people in urban
areas is astounding.   Many important quality events, from live
musical performances to the showing of classic films from other
cultures, need a shared taste by a minutely small percentage of
the population, sometimes 1 in 10,000 or less.   Even the location
where one enjoys this entertainment can be more personally
selected,   thanks   to   quality   home   stereophonic   systems   for
records, tapes, or video discs.   New digital technologies for
sound reproduction give ever greater possibilities for artistic
creation and listener enjoyment (Turner 1984, 47-48).   There is
also greater opportunity for being cosmopolitan, for book or
record   and   tape   stores   will   have   religious,   classical,   or
contemporary works from languages and cultures of all countries
and time periods.   Or we can get together in small, compatible
groups and sing, tell stories, or share ideas, as people do in
small communities and have done in other places and other times.
If people do less of some of these traditional activities, it is
because we have options to do others and not because we can't.
     One could go on indefinitely illustrating the decentraliza-
tion and enrichment of choice in the arts.   Scientific inquiry
is proceeding in the same direction, with wider availability
of books, computer home terminals, and the communication of
knowledge through telephone wires, microwave, or satellites. The
microchip and the transformation it engenders is truly a case
where small **is** beautiful.

Most of these decentralizing technologies are having their impact upon the lives and well-being of peoples in third world countries. Satellites for communication and ground imaging have their benefits in agriculture, public health, and many other aspects of third world economies. Many of the high-tech items for personal consumption and artistic satisfaction are in the far distant future of most third world countries. The primary relevance of their reference here is to counter the argument that using sophisticated technology ineluctably leads a country and people on a path toward centralization and alienation. It is made to counter an argument that, in our judgment, is without substance and is a strategy for slow economic growth and cultural stagnation.

In fairness reference must be made to a legitimate aspect of the argument for the other side, namely, that the decentralized technologies exalted here are the product of more centralized political and technological structures. The freedom that we have to travel about in automobiles or in airplanes implies the building of roads or airports and route structures. Even in democratic political systems people can feel that they do not effectively participate in the decisions as to where a new road or airport will be built and whose neighborhood will be disrupted. In sectors that are sometimes private and sometimes public (and sometimes bits of both), the cost of the technologies to produce a film, stage a play, or carry out an advanced scientific inquiry can mean decisions made by a few establish the framework for the range of choices by the many. That these problems are serious cannot be denied. It is one thing to recognize the problem and work to further democratize and increase participation in the more centralized decision; it is yet another to abandon the entire structure and its benefits to pursue the uncertain benefits of anarchic small scale development.

For all practical purposes, we live in an interdependent world. The term interdependent has become fashionable but is no less correct for being so. To say that we are all interdependent should not obscure the fact that some countries are more vulnerable than others. If we may be permitted to paraphrase Orwell, some countries are more interdependent than others. Being smaller, poorer, and late in arriving at more advanced technologies does make a country more vulnerable than are larger, richer countries with a large fund of knowledge and technological experience. To recognize the fact is neither to condone it nor to argue for alternate technologies. There is no viable autarchy option. Until we achieve a centralized international authority system (if we ever do) that can end warfare, countries will have to defend themselves. The country with technological advantage (either from within the country or brought from abroad) is more likely to be successful. Small industries will require at minimum a more sophisticated machine tool industry for parts replacement and repair. It looks increasingly evident that the most efficient and effective forms of solar power will be based on large-scale, sophisticated production of silicon cells. These are the minimum dependencies (or interdependencies) for a country attempting to follow an alternative technology policy.

If one starts with the proposition that being smaller, poorer, and less technologically advanced is inherently disadvantageous in terms of international economics, politics, and other relationships, then the question becomes one of which strategy more rapidly and effectively lessens this disadvantage. In our judgment, perpetuating technological backwardness is not a prescription for self-reliance. Even control of a "strategic resource" is not effective, as our chapter on minerals shows; science and technology can find ways around such strategic impasses. Essentially, a country must acquire the knowledge, skills, and other capabilities to take command of a technology and make it their own. This can be a long and arduous path, but it is the only way. For smaller countries, it may require some forms of regional cooperation. It is not a question of using indigenous technology, but of the indigenization of the very best in technology. Even if the country does not manufacture the technology in use, knowledge for operational command of the technology allows for effective bargaining among alternate private and public suppliers.

Technology is predicated upon knowledge of the physical principles of nature and command of the requisite means of gaining an understanding of these principles. Just as those who command a "natural resource" will find that others will use technology and science as a means of neutralizing its significance, so then in time will the gaining of knowledge allow those who follow to overcome the advantages of the pioneers of technology. The history of technology shows clearly that attempting to monopolize a technology is self-defeating in the long run. Far better it is to try to stay in the lead by advancing one's technology than by keeping it from others. However, in our contemporary world, where time is at a premium, the fact that attempting to monopolize a technology is self-defeating in the long run is of little comfort. It would be far better for all if we could develop effective means of fairly sharing the benefits, giving due reward to the innovators in all lands.

The problem-solving valuational process that is inherent in the use of technology involves an ethic that is antithetical to hierarchy and status. In the arts as well as in technology, the hierarchical belief system that elevated the mind and denigrated work with hands is counterproductive to the development of either. In ancient Greece, where metal workers were despised, the sciences of chemistry and metallurgy suffered. The arts, such as painting, advanced rapidly in fifteenth-century Europe, when it was recognized that they were "liberal arts" involving both a mental and physical activity (Brown 1980, 39). Technology has contributed to the advancement of scientific inquiry. Many modern achievements, from the horrifying unleashing of the power of the atom in nuclear bombs, to the space program, are falsely viewed as triumphs of science rather than of science, technology, and skills, of the mind and the hands working together. The hierarchical nature of German science in the 1930s and 1940s and the lack of integration with experimentation and engineering is seen as the reason why "the German scientists were unable to bridge the gap between theory and the completion of even the first step of building a working reactor" (Bernstein 1979, 68). In development, those who integrate technological knowledge with practices--the engineer who knows firsthand what equipment will do or not do, and the agricultural scientist who is willing to get dirt on his hands--make a dual contribution. They contribute to the advancement of technological capability and they advance an ethos that implicitly recognizes the integrative, cooperative relationship between thought, experimentation, and application among the people who perform these tasks. This is precisely the kind of cooperative social practice that the advocates of appropriate technology are trying to achieve. The hierarchical elitism that they rightly criticize is bad science, bad engineering, and counterproductive to economic and social development.

SCIENCE, TECHNOLOGY, AND IMAGINATION

This book argues throughout the economic benefits for countries that become participants in the dynamics of technology and science. This chapter focuses on the aesthetic and cultural benefits of this process. Beyond the practical achievements of feeding the hungry, curing the sick, and providing new technical means for the arts, science and technology have artistic values in and of themselves. In mathematics we speak of elegance, and in science of the beauty of theories, for they are works of the creative imagination of human beings. In our scientific inquiry we have learned what a brief flicker of time we have been here and what an incredibly small space we occupy. However, in this instant of time, how many barriers have we overcome to send people in machines soaring through the air daily, or breaking the bounds of gravity, by sending men and women in machines circling the globe or to the moon, with a frequency causing most of us to lose precise count of these momentous achievements. We send mechanical explorers through our solar system to send back images and other information, with one breaking free of our solar system to wander among the stars. Even so, we are still physically bound in time and space, but our imaginations are not. We are exploring a universe that ranges from 10^{25} meters (the distance between the farthest galaxies) to 10^{16} meters for quarks. There is speculation in carrying our understanding to 10^{-17} or even beyond to 10^{-31} or 10^{-32}. In time, our speculative inquiry carries from the first 10^{-43} second (Planck time) of the universe's existence to tens of billions of years hence. With science and technology we find regularity, order, and underlying principles in what is seemingly chaos. Even if we do not find regularities, we find ways of understanding chaos (or constrained randomness). When our theories are not fully satisfying at one level, we seek deeper structures and deeper understanding.

Somehow amidst all this there are those who argue that technology and science have dulled our imagination and creativity and turned us into automatons or adjuncts of the machines we serve.

Of modern inquiry one can say, as Darwin did of evolution, "there is grandeur to this view of life." For the universe, for life on our planet, and for human technology and culture, one can say as he did in The Origin of Species that "from so simple a beginning endless forms most beautiful and most wonderful have been and are being evolved." Intelligent concern, awareness of problems, enlightened policies that are truly economically, socially, and culturally developmental are essentials of any technology and science policy in third world countries. What is not needed are irrational fears of technology and science as allegedly dehumanizing activities. Ours has not been a Faustian bargain of trading our soul for knowledge, but rather, through knowledge we have gained greater insight into ourselves.

From earliest times there seems to have been a restless spirit in men seeking, striving, exploring, and conquering. Conquering too often has been of one's neighbor. In technology and science, this restless, unquenchable spirit is turned upon overcoming limitations to human development. In our time there is no more noble utilization of the powers of this spirit and the wondrous knowledge and technical instrumentalities it has created then turning it to the tasks of hunger and disease and other disabilities that stalk the planet. As we reach out with this global human enterprise of science and technology, sending out machines and messages to distant space, seeking out other intelligent life forms, we need to remember that there are real problems at home to solve if this enterprise is to continue. Messages from a lifeless planet are meaningless, as is human life without curiosity or concern beyond the span of one's immediate environment. There is nobility in reaching out to the stars and in feeding the hungry, and it is within the capability of science and technology to do both. Each helps to give meaning to the other. Though our knowledge of the immensity of the cosmos and its duration in time may make us seem small, it does not make humans seem insignificant. Curiosity and intelligence that have led to such understanding is precious and worthy of protection, and the opportunity for individual development rests in each and every instance of its existence.

PART III

Science, Technology, and Agriculture

CHAPTER 7

Nature and Technology

In his delightful book, A Choice of Catastrophes, Isaac
Asimov (1979b, 13) traces the etymology of the word catastrophe
to the Greek word meaning "to turn upside down." There is irony
here, since many of our contemporary catastrophists have turned
reality upside down in their predictions of cataclysms that have
yet to occur. Because of modern science and technology, we
live longer, healthier lives, yet somehow alternate technology
enthusiasts characterize these as death-dealing systems. These
enthusiasts keep talking about nature and natural phenomena, such
as natural food or the natural functions of the land. Taken
literally, such pronouncements are nonsense. Humans have always
been tool users, and this in itself gives us a different role in
the scheme of things. Malcolm Slesser (1976, 1) writes, "The
productivity of a natural eco-system is around 6 kilograms per
hectare per year of protein suitable for humans, and this is
only obtainable from the better land on the earth's surface.
Conceivably, such an unintensified system might support around
200 million people, a figure surpassed by the Middle Ages." Long
before the Middle Ages, of course, humans were transforming the
earth through agriculture and husbandry. As Rene Dubos (1980,
57) has noted, "Human beings have probably never been in real
'balance' with their environment except under conditions where
population density is extremely thin, as in polar regions or the
Australian desert." We don't live in a natural environment, and,
with over 4 billion people on earth, we cannot recreate one.

BACK TO NATURE?
Obviously, natural does not mean a literal "return to
nature," though one wonders why the phrase continues to be used.
All technologists--modern, alternative, or whatever--involve
humans living in an artificial environment. It is not the
naturalness that is important, but the quality and sustainability
of the life processes that can be carried out.
Small technology proponents place great emphasis on the
"redundancy" of simpler technological systems and the security
that this confers. A theoretical relationship between complexity
and reliability is assumed. Big, modern, and sophisticated

technology is fragile and subject to catastrophic disruptions. Such disruptions include power outages. Twice in over a decade the lights have gone out in New York City, and critics point to the vulnerability of modern technology. Presumably they take for granted the reliability of electrical systems that work for years with only occasional disruptions. What would the critics have us compare our electrical systems to for reliability--candles?

When it comes to the basic life-support system, it is clear that modern technology has created a redundancy never before achieved. In many instances, the reliability of modern technology is so high that reliability becomes a form of redundancy, providing a very high degree of security (Rosenberg 1982, 136). Basic to life support is the provision of food. Modern technology has regularized and stabilized food production, and this has been a factor in the steadily declining death rates that have accompanied the rise and spread of modern technology. Despite stabilization of output, the major variable for agriculture remains the weather. An agricultural system that is essentially worldwide provides greater security against the vagaries of weather through greater redundancy. Our modern transportation and distribution systems provide the means to alleviate severe need, the result being that "only a tenth as many people died of famine in the third quarter of the twentieth century as in the last quarter of the nineteenth century, despite the much larger population now" (Simon 1980, 1433).

Redundancy is at the very heart of modern technology. Hospitals have backup systems. Planes can fly even with the loss of an engine--or 2. Bridges are built with overload factors. And on it goes. If backup systems fail in a hospital, the disruption can be severe, and lives can be lost. But the severity of the disruption is a function of our temporary inability to derive the full benefit of a technology. If the technology were not inherently beneficial, its loss would be inconsequential. As we continually reiterate in our arguments with the small-is-beautiful enthusiasts, the aggregate of evidence of mortality attests to the redundancy (i.e., safety) of modern technology. With the spread of modern technology, death rates are falling almost everywhere. The very success of technology in protecting and sustaining life leads people like John Bryant (1980, 229) to perceive a need "to prevent the breeding of human beings who must have such technology to survive."

The human endeavor has always been subject to "severe disruption" by natural hazards such as earthquakes, floods, drought, and winds. Even here, however, technology provides protection. In a seminal study of natural disaster, Judith Dworkin (1974, 5) found that, with the exception of Japan, low death rates from disaster prevailed among high-income countries. Robert W. Kates (1978, 11) argues that "death rates from natural hazards in the United States may be down to the reasonably preventable annual minimum." Kates cites the standard figures on mortality from natural disasters: 1 in a million in the United States, 1 one in 100,000 in the world, 1 in 10,000 in Bangladesh.

It is true that most of the authors who write about risks of natural disasters also write of technological hazards. One such hazard stems from a willingness people have to live downstream from dams or in coastal areas subject to hurricanes. Our dams are so safe (not only in construction but also because of electronic sensing devices), and our hurricane warning systems are so effective, that some of us have been lulled into a false sense of security. But a developing country could, in fact, choose modern technology while simultaneously planning its settlement patterns to avoid risks. In already developed countries, undertaking resettlement would be costly. Technological hazards can be minimized by exercising common sense; natural hazards can be reduced primarily with modern technology. When we look at death rates from all causes, it is undeniable that modern technology has yielded a longer life expectancy.

By having modified nature in developed countries, modern technology has created the framework for the lifestyles that win the plaudits of nature lovers. Natural or organically grown food is not a prestige item in poor countries where dysentery is rampant, yet John Bryant (1980, 227) speaks about the "pollution of food." Natural childbirth means higher infant mortality except where technology has reduced the general level of disease. To poor farmers throughout history, "nature" has meant floods or droughts, locusts or ill winds. Modern technology, in tempering these hazards, has given the devotees of antitechnology (or at least antimodern technology) the opportunity to pursue their own courses. Rather than making each of us a mere cog in a machine, modern technology (particularly when coupled with democratic political institutions and traditions) has allowed more diversity and freedom of lifestyle than ever before. But small scale communities function insofar as people work together toward common goals. Choices are limited, so tastes must be similar.

Decentralization is another of the antitechnologists' proclaimed ideals. And it would be a worthy ideal—if valid. The irony is that our centralized, technologically sophisticated, democratic societies are functionally more decentralized than any others. Do-it-yourselfness is a function of the availability of modern technology in the home, and many of the do-it-yourselfers also benefit from the leisure afforded by modern technology. Ironically, centralized electric power, the great nemesis that the soft-energy advocates such as Amory Lovins wish to slay, has been a strong force for decentralization in the twentieth century. Because it can be transmitted and fractionalized, electric power has led to industrial decentralization (Rosenberg 1973, 77). Electrification has meant everything from home power tools to kitchen appliances to stereos to home computers; it has greatly expanded the range of individual choice for personal creativity, for both passive and active entertainment, and now even for home employment.

We can seek to improve our technology and make it less hazardous; we can strengthen the democratic processes in the collective decisions that provide the framework for our range of choices; we can work to give more people access to all the

benefits of modern technology. These goals are consistent with
trends already operative in modern technology. There are those
of us who like modern technology. If we unashamedly continue to
work to improve it, then achieving our objectives increases the
possibilities for the antitechnology people to achieve theirs.

FOOD PROCESSING AND HUMAN DEVELOPMENT
     Not only is "natural" not necessarily better, it has been
argued by some that food processing was historically a vital
part of human development. Cooking is considered a necessary
precondition for the human species to achieve "the evolutionary
superiority it holds today." KaKade and Liener write, "It was
only after man had learned to use fire for cooking, about 40,000
years ago, that it became possible for him to take advantage of a
greatly expanded food supply in the form of cooked vegetable
foods."
     As Rene Dubos has said, "Like it or not, from the moment we
learned to transform things according to functions we developed a
hundred thousand years ago, we drove the natural out" (Dubos and
Escande 1980, 99). Dubos characterizes the nature-knows-best
school of thought as a twentieth century version of Dr. Pangloss,
who thought we lived in the best of all possible worlds. This is
ironic, because critics of modern technology frequently refer to
its supporters as being Panglossian.
     Proper processing has been an integral part of various plant
and animal substances becoming food.

          Man's vegetable diet, without fire for cooking, is
     pretty much limited to special plant products like fruits
     and nuts. His digestive tract is simply not equipped to
     deal with cellulose and raw starch, which make up the bulk
     of vegetable material. The cellulose walls of plant cells
     are broken down by heat, and the starch is chemically
     changed into more digestible forms. Cooking, then, can be
     looked at as a sort of external, partial predigestion (Bates
     1967, 39).

Others are not readily digestible unless soaked in water; still
others need to have husks or shells or other parts removed. In
some instances, milling and reducing particle size facilitate
digestion. Heat, germination, fermentation--these can transform
and otherwise enhance the accessible nutrients of foodstuffs
(KaKade and Liener 1973, 232-38; Hulse 1982, 1291-92). There are
losses in food processing, but, as one group of nutritionalists
concluded, "some loss is inevitable, but for most, nutrient
losses are small" (Davidson et al. 1979, 213). They also find
advantages to processing, both in the processing itself and in
that, through storage and preservation, people have access to
foodstuffs at times of the year in which they are unavailable
fresh.
     An important coming together of technologies in food
processing was one that led to the creation of leavened wheat

bread. Not only did it require the original process of the
domestication of wheat, but it later resulted from "combining
three technologies: growing naked wheat, the invention of the
saddle stone, and the introduction of yeast" from wine-making
(Hall 1974, 137). Today in the United States stone-ground flour
has a prestige value implying a coarse grain size and improved
taste and nutrition. This is yet another instance of the
veneration of outmoded technologies without a basis in fact.
Steel rollers can grind flour to any particle size desired and
with no nutritional difference from that of the stone ground.
Preservatives, freezing, and other forms of protection for foods
can preserve nutritional qualities of foods, prevent spoilage
and other wastage, and protect against germ infestation of
potentially fatal consequences.

From 1940 to about 1978, more than 800 billion containers of
food were produced "with only 5 known deaths attributable to
botulism from that food" (Chou 1979, 20). This is an extraordinary, almost unbelievable achievement and shows that in one
critical respect our food is not polluted. There were 700 deaths
in the same time period from home-canned foods, which are
considered by some nature lovers to be more wholesome. Even
home canning today involves many twentieth-century household
technologies that likely make the product safer than preserved
foodstuffs of earlier centuries.

Those who in a blanket fashion condemn modern food
preservation must certainly be ignorant of the fact that foods
in history have been the carriers of botulism or ergot and
aflatoxins such as aspergillus, which have caused mass illness,
blindness, and, frequently, large-scale death. "Natural foods"
are not without their threats to life, as many writers have
noted. Because plants do not have the mobility of animals, they
are under "continuous evolutionary pressure to solve their
survival problems. This is frequently accomplished by chemical
means." According to Leopold and Ardrey (1972, 512), "Plants
accumulate many secondary substances--chemicals that do not
participate in the basic metabolism of the plant. Among these
are many chemicals that serve to repel or discourage the use of
the plant by insects, microorganisms, nematodes, grazing animals,
and man."

This is to be expected, since in the evolutionary scheme
of things, plant survival was the function of factors other
than serving the needs of humans (Boyer 1982, 443; Ames 1983,
1256). "Plants in nature synthesize toxic chemicals in large
amounts, apparently as a primary defense against the hordes of
bacterial, fungal, and insects and other animal predators,"
Ames says, "plants in the human diet are no exception." With
domestication, survival has depended upon the human need to
continue cultivation, but it does not mean toxity has been bred
out of plants. Undoubtedly, the yield, whether it be per land or
per labor unit, has been a primary consideration; nevertheless,
other considerations in selectivity make domesticated plant
evolution a complement to processing in making food accessible

to humans.   Such foods can in no meaningful way be called
"natural." Recent arguments for so-called "natural foods" are
that chemical additives cause cancer.   The term chemicals is
sometimes used alone as if plants themselves do not consist of
chemical constituents.   Ames argues that the major carcinogens
in what we eat are in the foods and not in the additives or
other "chemicals" in our environment: "Despite numerous
suggestions to the contrary, there is no convincing evidence of
any generalized increase in U.S. (or U.K.) cancer rates other
than what could plausibly be ascribed to the delayed effects of
previous increases in tobacco usage."   Further, "there are large
numbers of mutagens and carcinogens in every meal, all perfectly
natural and traditional.   Nature is not benign.   It should be
emphasized that no human diet can be entirely free of mutagens
and carcinogens" (p. 1261).   There is good news in Ames in that
some of the foods we eat also seem to provide protection against
cancer.
      It is ironic that people living in affluent regions
dominated by modern technology have the greatest opportunity of
enjoying small-scale technology and natural products with
relative impunity. So it is with foodstuffs. Colin Tudge (1980,
340) has estimated that a breast-feeding village woman (he used
one in Gambia as an example) would have to eat 2.8 kilograms of
unrefined plant food to obtain the approximately 2800 calories
needed to work and to lactate.   This is an "almost impossible"
task; for those who wish to be vegetarians, this is possible
because of "high energy plant foods obtained through milling,
through refinement (as in sugar) and through the oil press."   In
a word, the evolution of high-energy, high-protein plants,
coupled with food processing, has made the "natural" foods
lifestyle possible for the minority who care (and can afford) to
pursue it.
      The natural foods lifestyle in developed countries has
turned to organic foods. The term organic refers to raising food
crops without pesticides and without artificial fertilizers.
Tests in the United States have shown that many "organic" foods
have pesticide residues comparable to nonorganic foods (Delwiche
1970, 137-46; Consumer Reports 1980, 413-14).   The pesticides
could possibly have been in the soil from previous plantings
prior to a field becoming "organic," or from spraying on neigh-
boring fields.   The issue yet to be studied is whether the
successes of scattered organic farms around the country is
largely or at least in part a function of farming in a country
where most farmers actively engage in chemical pest control.
This is a critical question if these farming practices are to be
considered as prototypes for agriculture in other regions.
Further, the rapid growth in the United States and elsewhere of
no-tillage or minimal tillage agriculture, which preserves
groundwater and soils, requires herbicides and pesticides for
weed and pest control.
      One of the prevalent myths about organic foods is that they
derive their nutritive superiority from using manure and not

artificial fertilizers. "Actually, there is no scientifically proven difference in the nutrient content of plants that have been raised organically and those raised non-organically. Plants manufacture some of their own nutrients through built-in genetic programs," says Consumer Reports (1980, 413). However, as almost any introductory book on agriculture and soils will state, plants cannot directly use organic substances. Hartley (1976, 77) writes, "The protagonists of organic farming (who have many worthy motives for which we can have much sympathy) are, unintentionally, rather insulting to the clever green phoenix which is always rising from the ashes of the world. It alone does not need the products of organic chemistry; it makes them."

Organic material must be broken down by bacteria into its inorganic components before it can be absorbed and used by the plant. Broken down into components, the manure and artificial fertilizers are indistinguishable to the plant. There are 2 important differences. Artificial fertilizers can be composed of the right combination of nutrients for the soil in order for the crop to be grown. The nutrient contents of manures vary and do not always have the nutrients in correct proportions to meet the agricultural needs. Manure is generally considered better for the soil structure, though it poses other problems in that it can have toxic chemicals or a high salt content and may harbor bacteria, insects, worms and other pests (Hall 1974, 137; Consumer Reports 1980, 413).

SOLAR ENERGY AND PLANT GROWTH

The natural/organic food critics of modern technology in agriculture and other activities are generally part of the school of thought that argues that we are approaching the absolute physical limits, if we haven't already exceeded them, to the earth's capacity to sustain life. Continuation of our present path in using modern technology will lead to destruction of the soils, pollution of the atmosphere, and mass extinction of certain species. These views are critiqued generally in other chapters. This and subsequent chapters look at the human capability of using the earth to feed ourselves.

Several years ago, at the U.S. Agency for International Development, a speaker gave a delightful, witty lecture on the physical limits to growth, that was repeated on at least one occasion. The person who introduced him commented at the beginning of the question period that no one could possibly disagree and that the only purpose of questioning would be clarification. Complete with a chalk-drawn schematic diagram, the lecturer explained how sunlight fell on the earth and energy cycled through the system, and inevitably he invoked the principle of entropy. Several questions came into the irreverent listener's mind. First, entropy applies to a closed system, and the earth's system is an open one, since it derives energy from the sun. It is true that for any particular form of activity on earth or elsewhere, energy fed into a system emerges in a less usable form. Secondly, conceding that with current technology

the sun's energy does form an upper limit to agriculture and
other human life-sustaining activities, are we sufficiently close
to this limit that it should be an issue of policy concern for
economic development? Finally, given the principle of entropy,
how do alternate technologies and development strategies help us
to live within these limits, and do they help us stave off the
seemingly inevitable end, several billion years hence, to which
entropy seems to have destined us?

It can be argued that the earth is an open energy system
that takes energy from the sun, cycles some of it through the
system, and radiates into space a slightly greater amount. The
additional amount radiated back is from heat that escapes from
the interior of the earth. If the earth didn't radiate this
energy back into space, there would be a heat buildup that would
make life intolerable.

In fact, the Earth gains no net energy from the sun.
The crucial role played by the Sun for life on Earth, then,
is that it gives us high-grade energy in the form of
sunshine, and we return low-grade energy in the form of
infrared radiation. In the process of transforming the
former into the latter, a minute fraction of the free energy
is stored temporarily in plants, which in turn transfer it
(with inevitable degradation) to animals and to humans.
Occasionally, a human expends this free energy in the effort
of producing a symphony. Again, entropy--not energy--lies
at the root of the issue (Shu 1982, 80).

According to The Cambridge Encyclopedia of Earth Sciences,
"The Earth intercepts 1.72×10^{17} watts of solar energy (5.42×10^{24} J/a), of which about 35 percent is reflected by clouds and
land surfaces or is back-scattered by dust particles in the upper
atmosphere, and 65 percent is absorbed in the atmosphere or at
the Earth's surface." For each hectare of land in the temperate
zone, the solar energy ranges from $15-40 \times 10^{6}$ kcal; the total
per hectare per year ranges from $1.1 \times 1.8 \times 10^{10}$ kcal with 1.4×10^{10} kcal as a reliable average" (Pimentel and Pimental 1979,
13). "Annually, the total light energy fixed by green plants in
ecosystems is estimated at about 400×10^{15} kcal divided equally
between terrestrial and ocean systems. . . . [plants fix] less
than 0.1% of the total sunlight reaching the earth. . . . In
agricultural ecosystems, an estimated 15×10^{6} kcal of light
energy (net reproduction) is fixed per hectare per crop season,"
which again is about 0.1% of that available during the year.
Another author estimates that "only about one part in 200,000 of
the sunlight falling on the earth is converted into food energy
for human beings" (Revelle 1976, 166). It would appear that at
present it is not a significant limiting factor for plant life
and/or agriculture.

Plants use only a small fraction of the sunlight they
receive. Of the sunlight received, approximately half is in the
spectrum unusable by plants. Another 12.5-13% is reflected back
through the leaves (Pirie 1976, 130). A large portion of the

rest is used for performing vital plant functions, such as transporting water and nutrients. The National Research Council (1977, 76) suggests a theoretical maximum capture and conversion of 12%--this contrasts with the 1-3% characteristics of most plants. Rather than being a limiting factor, more effective use of the sun's energy is seen by many as a research potential for increasing agricultural output, particularly with respect to genetic and chemical control of photorespiration. Wittwer ([1977] 1978, 487-95) defines the main items of this research agenda:

> They include identification and control of the mechanisms that regulate and could reduce the wasteful processes of both dark and light induced (photo) respiratory mechanisms responsible for redistribution of photosynthates which in turn regulate yield and maximize the Harvest Index; resolution of the hormonal mechanisms that control flowering and leaf senescence; improvements in plant architecture and anatomy, cropping systems, planting designs and cultural practices for better light reception; and carbon dioxide enrichment of crop atmosphere.

Energy, of course, is only one part of plant and animal growth. Oxygen, carbon, hydrogen, nitrogen, and a host of other elements are needed to create living matter. For the main 4 plus most others, life on earth uses only a small fraction of the elements potentially available. Vajk (1978, 64) writes, "The amount of each of the four principle elements of the biomass of the planet which are available in the atmosphere, the oceans and in oil shale or limestone deposits is in each case one million times greater than the mass of that element in the biomass of the planet." Of these elements in atmospheric forms that are directly usable, in the case of oxygen only one part in 5,000 is used, and for carbon, "three parts in 10,000 of the atmospheric carbon dioxide are utilized" (Revelle 1976, 166). In both cases they are returned to the atmosphere. Of the minor trace elements necessary for life, most of these are abundant. The most serious question concerns the availability of arable land and water. The superabundance of all factors for life would count for nought if one or more essential items were missing. In economics we can substitute one factor for another. In life processes the issue is not so simple, and limiting factors tend to be more restrictive.

THE CONCEPT OF LAND

The classical economists following Ricardo spoke of land as "the original and indestructable powers of the soil." In micro-economics we used to define it for students as the "nonhuman, nonmanmade factor of production." Economists never rigidly adhered to these definitions, but they certainly provided the framework for conceptualizing about land. Resources would also fit at least the second of these 2 definitions, in which was implied the idea that they were in some sense natural. Because

land, soil, and resources were natural, they were fixed and
finite. It was recognized by Ricardo and others that, although
land was fixed, output could be increased by more intensive
cultivation. However, in Ricardo not only did rents rise, but
output per unit of input fell. Thus, the idea of diminishing
returns (now referred to as variable proportions) meant
continuing to add a variable factor of production to a fixed one.
Ricardo believed that rising rents and falling profits would lead
eventually to a rate of return in which capital accumulation
would cease and growth rates would be zero. This was called
the "stationary state." With no increase in subsistence, there
would be no increase in population. To Ricardo's rival, Thomas
Malthus, there were no such automatic limiting factors short of
catastrophe. People could undertake measures such as abstinence
and late marriages ("moral restraint") to prevent the otherwise
inevitable. The principle (note the singular) of population was
geometrically increasing population while food supplies increased
arithmetically.

We can take a different approach and understand land in much
the same theoretical framework that we used for mineral (i.e.,
"natural") resources. The resource character of the raw stuff of
the universe is not a property of the substance but of human
technology and the ability to use that technology for human
purposes. A similar analysis of soils was made by the geographer
Preston James (quoted in DeGregori 1969, 47). He notes that
"fertility is not a quality inherent in the soil alone; it can
only be measured in terms of specific soil uses." For that 99%
of human existence when we were hunters and gatherers, fertile
soil (or good land) was that which supported accessible, edible
flora and fauna. Even in this period, as we have noted, changes
in technology were altering the resource character of the land.
The stone tools that allowed humans to harvest roots from the
rock-hard soils of the East African savanna were increasing
the fertility of the soil (from the human perspective, of
course). They were also, in a small way, inaugurating the human
transformation of the landscape. When, with tools and social
organization, humans would hunt large animals for food (and for
other resources, such as skins and bones), then the land that
supported these animals became more fertile for human purposes.

Along with the domestication of plants and animals, humans
began more consciously to change the land. Many of the river
bottoms into which humans moved in large numbers were malarial
and waterlogged. The marshes and swamps were drained, the land
was tamed, and the regimen of the river was used for human
purposes such as irrigation and transportation. Among these were
the rivers, such as the Nile, that we identify with the great
civilizations of the past. Agriculture created arable land.
Changing technologies have continued to change our concept and
perception of land and its fertility. People moving to a new
area tend to take their technology (and, to the degree possible,
their institutions and belief systems) with them. When the new
lands are not productive, the fault is attributed to the inherent

lack of fertility of the soil and to the climate. There is a
massive amount of literature on the adaptations and developments
that made such lands into the most productive in the world.
Technological diffusion in agriculture, like diffusion in other
areas, generally involves a combination of modification and
adaptation of the imported technology, some indigenous invention,
and some items that could be borrowed without modification. The
successful integration of these three components of technology
transfer has transformed historically infertile soils and
climates into productive lands.

We could give many other episodes where a land was believed
unproductive because the wrong technology was applied to it.
Ruttan (1983, 11) writes, "In the developed countries agriculture
has made a transition from a resource-based to science-based
industry. In 1925 corn yields in Argentina were higher than
those in the United States. Fifty years later corn yields were
more than twice as high in the United States as they were in
Argentina." Argentina's yields in 1925 were the result of
technological changes that made the renowned Pampas productive
for cattle, grains, and other forms of agriculture. Some of the
most productive cereal-growing areas of the United States are
part of an area that was called the Great American Desert (Webb
1931).

Economists such as Theodore Schultz (1977, 1981) have long
argued that some of today's best agricultural lands once had poor
soils. Schultz says "the original soils of western Europe,
except for the Po Valley and some parts of England and France,
were in general very poor in quality. As farmland, these soils
are now highly productive." Income and output differences
among agricultural peoples are not explicable by differential
soil productivity. To Schultz, "a substantial part of the
productivity of farmland is man-made by investments in land
improvements." A similar view is held by N. W. Pirie ([1976]
1981, 305), namely that "good farmland is usually created by
skilled farming."

The process of arable land creation continues to the
present. Dubos (1980, 90) tells how "in Australia the soil of
the 90-Mile Desert east of Adelaide contained virtually no
phosphorus, copper, zinc, or nitrogen and supported only a
miserable vegetation. During the 1940s these missing elements
were added to the soil . . . [and in] 20 years, the low-nutrient
desert had become a rich pastureland." He goes on to note the
deserts that have been made to bloom (de-desertification) and the
"unproductive heaths" of Europe that have been transformed and
made productive. Also, of course, he mentions the historic
process of land creation from the sea in the Netherlands, where 6
out of 10 people live below sea level.

Julian Simon (1980, 1981, 86) is one of the most outspoken
proponents of the idea that arable land is created by humans. In
a controversial book, The Ultimate Resource, as well as in his
earlier article, "Resources, Population, Environment," he states,
"The key idea is that land is man-made like other inputs to farm

production."   He argues that both in the United States and
worldwide new land is being created faster than it is being lost
to highways, cities, or desertification.   Land also is being
taken out of production in the United States to revert it to
forest, but this, argues Simon, is because increased output in
other lands has rendered it unneeded. This is also the result of
mechanization placing a premium on flat land, causing hilly
terrain to be withdrawn (pp. 81-89).   The data Simon uses on
arable land were strongly challenged by letters to Science (1980,
1296-1308) critical of his article.   One very good point that
several letter writers made is that aggregate data on hectares of
arable land do not take into consideration the fact that
population growth may be forcing people to bring certain areas
into cultivation that cannot sustain it.

There are many more ways of creating new land.   Bringing
uncultivated land into production is only one.   Not only can
irrigation bring cultivation to arid lands, it can frequently add
a second or third crop season to the agriculture of a region.
Breeding crops with shorter growing seasons that allow for an
extra crop is yet another.   Research has been carried out and is
continuing in finding or breeding crops that are more tolerant of
salt or can withstand the stress of acidity from aluminum-toxic
soils.   The National Research Council (1977, 79) says "about 40
percent of the world's potentially arable soils, more than a
billion hectares, are acidic types that contain soluble aluminum
(or related compounds) to restrict growth."   Some of these lands
have already been brought into production, particularly in the
Brazilian Campo Cerrado.   Only about 10 million hectares out of
180 million in the Campo Cerrado are currently in production.
Anywhere from 50 million to 80 million hectares are cultivable
with proper application of currently available technology.
Triticale and new varieties of sorghum and soybeans are among the
crops that have potential for higher aluminum tolerance.   Beets
are one of the potential halophytes (salt-tolerant plants).   If,
or more accurately when, the appropriate crops are bred, then in
a very real sense scientific research in plant breeding will have
created arable land.

Even without scientific research and new crops, it still
is not clear whether we are at the limits of cultivation and
food supply.   The Buringh et al. (1975, 59) study concluded
that the total potential land for agriculture was 3,419 million
hectares, of which 1,406 million are currently under cultivation.
Even allowing for the huge costs for irrigation, reclamation,
fertilizer, etc., the data still indicate considerable potential
for growth.

The problem of arable land creation for third world
countries is that much of it is largely dependent upon the spread
of modern technology and agricultural industries (Kellogg 1973,
84).   Kellogg estimates--obviously for the time of publication in
1973--that the "total area of potentially arable soils is more
than double the total now being farmed."   In 1980, another
estimate was "that land available for crops through future

irrigation is 1.1 billion hectares, or enough to feed more than
10 billion people at twice the FAO levels" (Scrimshaw and Taylor
1980, 83).   It was this technology that created many of the
productive lands of the industrial world.   A number of authors
(Kellogg 1973, for example) have noted that the soils in Florida
in the United States were not very productive in "their natural
condition."

     New systems of continuous cultivation offer possibilities of
agriculture   in   the   humid   tropics   without   ecological   damage
(Sanchez et al. 1982, 821-27; Nicholaides et al. 1983, 119,
125-33).   Possibly as many as 200 million hectares out of 484
million hectares of land in the Amazon basin are cultivable
(Nicholaides et al. 1983, 139).   Margaret Mead said, "The future
of mankind is open-ended."   Theodore Schultz (1981, 6), in
quoting Mead, adds, "Mankind's future is not foreordained by
space, energy, and cropland.   It will be determined by the
intelligent evolution of humanity."

SOIL EROSION AND CONSERVATION
     The   literature   on   soil   exhaustion,   soil   erosion,
desertification, and other perils to agriculture is vast and
growing.   There are frequent short newspaper stories and
occasionally longer stories or series.   There are popular and
scholarly books and articles.   The tone is sometimes shrill.   As
a previous chapter indicates, the track record of some of the
extreme catastrophists leaves much to be desired.   Some have made
very specific predictions that have failed, fortunately, to come
about.   On the other side, the most protechnology optimists
basically recognize that we have problems.   Even Julian Simon
(1981, 89), who is generally recognized as being at one extreme,
makes it clear that his "message is not one of complacency."
It is not that we do not have problems of world food supply,
population growth, desertification, habitat destruction, and
species extinction.   The real issue is the nature of the problem.
Problems must be understood before being solved.   It is one thing
for writers in developed countries to scream about the need for
world population control, and it is another to come up with
means to achieve it.   Ironically, it seems that where modern
technology, which is often viewed as the culprit, has been able
to facilitate rapid rises in incomes, birthrates tend to follow
death rates downward.   Birthrates eventually begin falling at a
faster rate, lowering population growth and beginning what has
been called the demographic transition.   This has already
happened worldwide.   Unfortunately, in some parts of the world,
notably most of sub-Saharan Africa, population growth rates
remain high while food production per capita is falling.   Since
many governments in these regions, along with international
and donor country aid organizations, research institutes, and
scholars throughout the world, are working on the problem, we are
not in need of hysterical articles calling attention to it.
     On food supply, the scholarly sources cited indicate that
absolute physical limitations are not the cause of food supply

problems. In many instances food is available. Those who need
it lack the skills, capital, opportunities, etc., to acquire the
income to obtain it. In other instances potentially arable
land may be there, but the capital for roads, irrigation,
fertilization, and other improvements may not. Too often the
problem is further compounded by government pricing policies that
discourage production. Scientific research and new technology
will add in the coming years to the stock of arable land, but
this does not mean that it will be brought into production even
if the output is needed.

Treating land in the same analytical framework in which we
studied mineral resources can give us a different perspective
on the issue. As noted, using resources is not the same as
using them up. If the society that uses minerals uses them
intelligently by continuing exploration, advancing knowledge,
and not deliberately engaging in waste, then we can continue
indefinitely to have minerals from the earth and, eventually,
from the heavens. Similarly, if we use our soils and practice
the latest in agricultural technology and scientific research,
both for production and for conservation, then arable land and
agricultural output can continue to expand for some time to come.
This certainly provides the time and the means to address the
population problem. The physical limits to production are not
the problem in the present; technological and capital and even
policy limits are. Modern technology is not the problem; it is
the most effective means for the solution.

One of the significant studies on soil loss worldwide is
Erik Eckholm's Losing Ground (1976). Eckholm draws heavily from
the experience of Nepal, where the mountain forests are being cut
for fuel and to create land for cultivation. Within a few years
the land is totally eroded and useless for life sustaining
purposes of any kind. The lesson would seem to be that a static
technology (in this case both in fuel and agriculture), coupled
with a growing population, is destructive of the environment,
sometimes irreversibly. If agriculture in other areas of the
country were more productive, and were there other fuels and
employment opportunities, the situation would be considerably
different. Similarly, throughout the tropics people are moving
on to marginal lands because of inadequate food and/or employment
opportunities. In many cases they are moving onto lands where,
as noted above, it would be possible to have a sustainable
regimen of agriculture.

It is in the humid tropics that we have the most egregious
and catastrophic myths about agriculture. This is an ironic
twist from the earlier mythology of lushness. The fragility of
some ecosystems and the poor quality of some of the soils is
generalized to all of them. Careful empirical research of
regions, however, such as those already cited on the Amazon basin
(Sanchez et al. 1982; Moran 1983, 308; Nicholaides et al. 1983,
115), find a considerable diversity in the quality of the
tropical soils and the potential for development. Nicholaides et
al. write, "The old laterization myth that the Amazon Basin soils

will turn to brick when cleaned is just that--a myth." As Moran
states it, "recent research has balanced our views of soils
of the Amazon. They are recognized by a growing number of
persons as highly variable in quality and with a broad range of
potential. Some indeed live up to the stereotype mentioned,
whereas others can be highly productive over many years. The
problem is developing proper management systems for each of the
soils." Proper use and treatment of these tropical soils can
enhance them and create economic value in the same manner that
arable land has been created for other soils. According to
Nicholaides et al. (p. 133), "Appropriate fertilization and
continuous cultivation after seven years have improved rather
than degraded" the soils of the upper Amazon at a project in
Yurimaquas, Peru.

Thus, the belief that tropical soils quickly become depleted
through uses other than their natural use (such as rain forests)
may be true for some, but not for all. For the others it is a
function of the technology of agriculture, not the inherent
properties of the soil. Some of the dominant soil types of the
tropics are similar to those of the eastern United States
(particularly southeastern). These are Ultisols (formerly
designated red and yellow podozolized laterites) and Oxisols
(true laterites and latosols), which constitute "about 70% of the
unused tropical land." It is estimated that they "can be chem-
ically amended to permit acid-tolerant crop selection to produce
80% of U.S. yield . . . [and] 150-200% of that in temperate zones
for annual yields with multicropping" (Christianson 1982, 4).
It is interesting that some worry about the "fragility" of the
tropical rain forest ecosystem because it is complex, while
others worry about the "fragility" of modern agricultural
ecosystems because they are simple. Since over 90% of the foodstuffs that humans consume are probably from plants domesticated
in tropical or subtropical areas, we ought to be able to find an
effective way to grow some of them there. Unfortunately, the
poverty that leads to cultivation of these lands rarely equips
people with the means for any but environmentally destructive
agricultural practices.

Clearly, the experience in the United States shows that, as
agricultural land becomes more productive, some of it gets taken
out of production and is allowed to revert to forest or pasture.
If in the United States soil is being lost in the cultivated
areas, it is not because modern science and technology have
not given us the means for ecologically sound, sustainable
agriculture. Though it is a myth that we are running out of
cropland in the United States, in some areas we may be ruining
the soil and drawing down the water table. Again, the causes of
this are complex--i.e., social, political, economic, and legal--
and not technological.

Humans are unique in that they are not bound to an original
environmental or ecological niche; instead they have been able to
survive in many different areas and create the conditions for
their existence. Humans are the only mammals that spread across

the entire globe without speciation. Dubos writes, "The most
powerful animals in the world never leave the restricted
ecological zone to which they've adapted. . . . But for the past
ten thousand years, and perhaps before that, men have been
transforming the land in order to adapt it to themselves. And
that, I believe, constitutes a most fundamental difference
setting the human race apart from all other species" (Dubos and
Escande 1980, 100).

There are well over 4.5 billion of us on this earth. Never
before have so many lived in affluence or at least had such a
high degree of food security. Population growth means that
it may also be true that never have so many lived in poverty
and food insecurity. The numbers of those receiving less
than the critical minimum of food varies greatly according to
the source consulted (Mellor and Johnston 1984, 32). However,
as a portion of the total, the size of this latter group has
generally been falling since World War II. Like it or not, if
we are to sustain food security for those who have it, and
extend this benefit to those who have not, then we have little
choice but to continue on a path of scientific and technological
research and implementation. Nature will not provide, nor will
agricultural techniques if we destroy the land and its cover.
Technology as problem solving involves human ingenuity that
creates land and resources and turns the power and principles of
nature to human purposes. This endeavor is what Dubos called
"the humanizing of the environment."

CHAPTER 8
Soils, Chemistry, and Water

The previous chapter argued that sunlight and land surface are not factors that will limit agricultural production in the foreseeable future. Sunlight is certainly the closest approximation to a wholly "natural resource." Land surface, as well, is largely a "natural resource," although, as noted, productive lands have been reclaimed from the sea. When we go from the concept of land surface to arable land or potentially arable land, however, then we are talking about human investment, ingenuity, and technology. With current technology, we argue, the arable and potentially arable lands are more than sufficient to accommodate considerable population growth and allow for continued increases in per capita food consumption. The concept of arable and potentially arable land involves other assumptions. Potentially arable land may be potential because it requires investment for draining or some form of rehabilitation. Both arable and potentially arable land can be deficient in vital building blocks for agricultural plant life.

SOILS AND PLANT NUTRIENT NEEDS
Plants, like other forms of life, require certain specific chemical constituents plus energy to build the basic organic molecules. In the study of soils and agriculture, the work by Justus von Liebig in the nineteenth century recognized that plants are limited in their ability to build their physical life structure by the least available component (Janick et al. 1969, 218; Binswanger and Ruttan 1978, 372).
Not all plant needs are the same. The plants that became the basic material for domestication evolved in response to specific conditions of their habitat. There are an array of different soils in the world with differing mineral components, and most of them support (or supported at one time) an indigenous plant life. The vast majority of plant life still takes place on uncultivated land. "The total dry matter produced by photosynthesis is a massive 116 thousand million tonnes, the energy equivalent of six times the world's annual consumption of oil. But only a small fraction (0.8%) takes place on cropland" (Food and Agriculture Organization 1982). With domestication, agri-

culturalists used plants that were adapted or adaptable to the particular soils, rainfall, temperature, and so on, of the land they were cultivating. They also modified the land to fit their agricultural needs. Despite wide variation in specific plant needs, all plants require light, water, nutrients, and heat for growth and reproduction.

Plant nutrients can be considered in 3 categories. They are major elements, or macronutrients; secondary elements; and micronutrients. The categories are purely quantitative, since all are essential in varying amounts (Sprague 1973, 93). Nitrogen, potassium, and phosphorus are the three major elements from the soil necessary for plant growth, and, as would be expected, they are the major components of the world's fertilizers. Nitrogen may originate in the atmosphere, but it is only available for plant use when it is fixed in the soil or on the root, and occasionally on the leaf. Nitrogen is "fixed" by turning it into a form such as ammonia (NH_3) that is assimilable by the plant. Carbon, hydrogen, and oxygen are the basic components of carbohydrates that are formed through photosynthesis from carbon dioxide and water ($H_2O + CO_2$ + light energy ———> $-CHOH + O_2$). Derived from the air and from the soil, they constitute over 99% of the mass of all living matter. Phosphorus is necessary for both initiating plant growth and bringing seeds to maturity for reproduction, and it is necessary for the production of DNA. Nitrogen is necessary for the creation of proteins and nucleic acids. Potassium is required for the production of sugars, starch, protein, cell division, and growth. Calcium, sulfur, magnesium, and chlorine are next in quantitative importance. The micronutrients, or trace elements, are iron, copper, manganese, boron, zinc, and molybdenum. The requirements for plant growth will differ according to the soils, the particular plants, and the nutrients they need. Some indication of the range can be shown by the fact that "about 50 to 100 pounds of supplemented nitrogen may be required for optimum growth, whereas an ounce of molybdenum may be adequate. Inadequate amounts of either will seriously limit yields" (Sprague 1973, 93).

The processes by which nutrients enter living matter, work their way through the foodchain, return to the environment and then back into living matter, are referred to as cycles. The most famous of these cycles, which are included in most basic texts in geography and life sciences, are the carbon cycle and the nitrogen cycle. Both of these have a gaseous, atmospheric phase that widely distributes them to where, by proximity, they are at least serviceable. Both living matter and the atmosphere constitute a large storehouse of both. For carbon, photosynthesis and respiration cycle a significant part of this total through the system each year.

One of the most important cycles is the phosphorus cycle. Essentially, phosphorus does not exist in gaseous form, though it is present in minute quantities in atmospheric dust. The various processes in the phosphorus cycle are slow and the redistribution is uneven except within fairly stable ecosystems. One of the first fertilizers (other than manure) consisted of bones broken

up and added to the soil. Many phosphoric ores are the products
of sediments from marine invertebrates. Bat guano and bird
dung (from birds that feed on ocean life) are also a major
source of phosphorus. It has been suggested by some (not
necessarily catastrophists) that in the future we will have to
mine graveyards for sources of phosphorus for agriculture. In
terms of crustal abundance and accessibility, phosphorus is the
macronutrient most likely to be in short supply.

The quantitative significance of soil nutrients is not
totally reflected in the proportion of the biomass resulting from
photosynthesis and plant growth and the rest of the food chain
derived from them. E. S. Deevey (1970, 148-58) has estimated the
proportion of each of the 15 most abundant nutrients in the
global total of living matter.

The three largest components of living matter are the basic
carbohydrates, and they are

Hydrogen	49.74
Carbon	24.90
Oxygen	24.83
Subtotal	99.47

The other nutrients are

Nitrogen	0.272
Calcium	0.072
Potassium	0.044
Silicon	0.033
Magnesium	0.031
Sulfur	0.017
Aluminum	0.016
Phosphorus	0.013
Chlorine	0.011
Sodium	0.006
Iron	0.005
Manganese	0.003

There are many factors that can cause land to be currently
nonarable. Nonarable land can include land that supports
vegetation but that cannot sustain domesticated cultigens. Some
lands are potentially arable but require large investments in
irrigation or drainage, the addition of soil nutriments, or the
removal of salts or toxic substances. Africa, and to a lesser
degree South America, have an abundance of such soils but are
critically short of the capital for the investments necessary to
bring this land to a condition for sustainable cultivation. Even
Asia has arable land that is not now being cultivated. In
addition, there are, as already noted, large areas of land that
are not defined as arable with current technology, but for which
research is underway for new crops or agricultural techniques
that will allow them to be cultivated at a viable investment
cost.

SOIL STRUCTURE

According to Brady (1982, 848), "Among the most critical constraints on production in developing countries is that of problem soils." Soil acidity is a major problem, particularly in the tropics where the soils have been leached. The pH of the soil is an indicator of the free hydrogen ions in the soil and therefore is a measure of acidity and/or alkalinity. The pH is a term in chemistry that refers to the negative logarithm of the hydrogen ion concentration. The scale is 0 to 14 with 7 being neutral (10^{-7} mole per liter) in pure water. The difference in hydrogen concentration between each whole unit is a factor of 10. Above 7 indicates alkalinity; below 7 denotes acidity. Most soils register somwhere between 3 and 10. The pH is an important determinant of the availability of the soil's nutriment for a plant. Extremes of acidity or alkalinity may actually facilitate the availability of some nutrients. The superabundance of some nutrients is less important than the growth limitations placed upon plants by the shortage of others. The nutrient demands of the plant will determine the range of acidic and alkaline soils in which the cultigen can be grown and the specific type of soil, acidic or alkaline, that is conducive to optimal growth. Adding lime is the way in which the acidity of the soil is reduced.

The chemical fertility of soils is focused on the clay particles. Clays consist of small particles of mineral and organic origin that are suspended as colloids in the soil moisture. These colloids are negatively charged and attract positively charged ions called cations. These cations replace each other through time in ordered sequence from sodium to potassium, magnesium, calcium, and finally hydroxyl aluminum. In the soils of humid climates, most profoundly in the warm humid (tropical) climates, this cation replacement process allows the base cations (potassium, magnesium, calcium, and sodium) to be lost to the soil through leaching. As a result most soils of humid climates, both cool and warm, are acidic. This ultimate replacement of the base cations by hydroxyl aluminum is referred to as aluminum toxicity and is characteristic of many tropical soils that are not now arable because of a low cation exchange capacity (CEC). Typically, the CEC may be increased by the addition of lime (CaO or $CaCO_3$), followed by the application of the essential fertilizer minerals: nitrates, phosphates, and potash (potassium).

Beyond the mechanical functions such as draining lands, leveling, terracing, or removing rocks, etc., adding chemicals to the soil has been and remains the primary means by which arable land is created and sustained. Prior to human occupation of land, the mineral or nutrient cycle for the land was relatively closed. Plants began, grew and died, and their nutrients were returned to the soil. Animals used plant material and their waste, and their carcasses were returned to the land in the same general vicinity. The water cycle was more complex, involving runoff or evaporation and its return as rain. Microorganisms, some associated directly with plant life, fixed nitrogen in the soil, which involved reducing it or hydrogenating it, thereby adding electrons to it.   Other microorganisms oxidized (or removed electrons) and volitilized nitrogen.   Some of the other nutrient cycles are more complex and for some, such as sulfur, the mechanism by which sufficient quantities are available for plant life is still not fully understood (Deevey 1970, 148-58). One need not make claims of the harmony of nature or assert other virtues to recognize that there were sustaining processes generally in operation.

When humans create arable land or begin cultivating it, they disrupt these nutrients and whatever balance exists in natural processes.   As one author (Hartley 1976, 73) states, "Agriculture does not disturb this balance; it destroys and replaces it." Quoting Sir Vincent Wigglesworth, G. S. Hartley further defines agriculture as "the art of disturbing the balance of nature most safely to our advantage."   All agriculture, certainly all agri-culture beyond subsistence, is essentially an export activity. The nutrient cycle is disrupted as nutrients that were taken from the soil by the plants are transported to a different locality, where as waste they are concentrated.   Unless these animal and vegetable wastes are consciously returned in appropriate quantities, the regimen of agriculture depletes the soil of nutrients.

The creation of arable land at times involves a process as simple as adding lime to the soil to reduce acidity.   This adds calcium, though there is danger that adding too much calcium (or adding it at the wrong time) can bind other nutrients and make them unavailable for plant use.   Adding human or animal waste is yet another simple means of bringing or returning nutrients to the soil.   In the nineteenth century, as the study of soils became more scientific, other means of sustaining or enhancing soil fertility were devised.   Drawing on accumulated wastes, guano from bats and mineral deposits such as nitrates, nutrients could be brought to soils from around the world.   The main nutrients thus derived from mineral sources were nitrogen, phosphorus, and potassium.   The transporting of soil nutrients across oceans not only helped sustain agriculture in already settled areas, but it also helped to expand and intensify the long-established process of creating arable land.   Bringing dry land into cultivation through irrigation, adding an extra crop season in an agricultural cycle (thereby increasing the crop intensity of the land), or reducing the fallow period or time for indigenous vegetation regeneration, are all ways of creating "new" land.   All require the nutrients to be brought in from other sources.

FERTILIZATION AND NITROGEN REQUIREMENTS

Prior to the twentieth century, other than human and animal wastes, green manure, and the naturally recycled materials, nutrients were mined and only minimally processed.   Even in composting, the transformation of vegetable matter was done naturally by microorganisms, with minimal human intervention.   In

this century humans began the process of manufacturing soil
nutrients with the development of industrial fixation of
atmospheric nitrogen by Fritz Haber and Karl Bosch. With fossil
fuels and hydrogen as a heat source and iron as a catalyst,
nitrogen was combined with hydrogen to form ammonia (NH_3). Urea
and nitrates, two nitrogen compounds also useful as fertilizer,
can be made from ammonia (Brill 1977, 68-69). Following World
War II there was an enormous growth in the production of
synthetic nitrogen fertilizer (Delwiche 1970, 137). The period
from World War II to 1973 was a period of declining real energy
prices and greatly enhanced supply, both of which facilitated the
expansion of fertilizer production. Agriculture in developed and
developing countries responded to the increasing availability of
fertilizer. More and more was used to increase output. Crops
were bred that could respond to increased amounts of fertilizer
without lodging. Crops also were bred for shorter growing
seasons, thus allowing for more crops and requiring more soil
nutrients. Increases in irrigation also allowed for more crops
per year. This process--improved seeds, fertilizer, and so on--
brought great increases in world per capita food supplies.

Since 1973 increasing energy prices have raised the cost of
manufacturing nitrogen fertilizer. Many of the other nutrients,
such as phosphorus, are not as energy intensive in processing,
but they do require costly energy to transport them (Sheldon
1982, 49-50). Interest has turned increasingly to other nutrient
sources, such as nitrogen-fixing bacteria and more efficient use
of fertilizers. "Inefficient crop use of applied and soil bound
nutrients is a serious constraint to production in developing
countries," writes Brady (1982, 849). "The enormous losses of
applied nitrogen provide perhaps the most significant challenges
to chemists and biologists alike."

To Brady, more efficient use of nitrogen fertilizer is
critical in order to increase or even maintain future food
production. He refers to the "notoriously low efficiency with
which plants utilize inorganic nitrogen, especially that added in
fertilizers." Brady indicates that some crops recover 50-60% of
the nitrogen, while others are as low as 25-35%. The bulk of the
loss of nitrogen results either from leaching or volatilization.

The ammonium ion (NH_4^+) can be oxidized (nitrified) to
nitrite (NO_2^-) and/or nitrate (NO_3^-). These in turn can be
reduced (denitrified) to nitrogen (N_2) or nitrous oxide (N_2O)
(Delwiche 1970). In denitrification the nitrogen is lost to the
atmosphere through volatilization. When nitrified, the nitrogen
is more readily lost through leaching. Nitrogen also can be lost
by irreversible binding of their compounds to organic matter in
the soil (Brady 1982).

Brady writes that, "attempts have been made to control
nitrogen loss by (1) developing slowly available nitrogen
compounds, (2) encapsulating or coating conventional fertilizers
with a chemical that reduces the rate of nitrogen availability,
and (3) nitrification inhibitors that keep the nitrogen in the
ammonium form, thereby reducing the risk of its being lost by
leaching or volatization."

Technically these 3 means for controlling nitrogen loss are feasible. There are several nitrogen compounds that release nitrogen slowly into the soil. Urea can be coated with sulfur to control its release and there are a variety of chemicals that serve as nitrification inhibitors. The problem with all three of these means of controlling nitrogen loss is the cost. In some circumstances these techniques are economically advantageous. Clearly, there are technical means to solve the problem of a more effective use of nitrogen; there needs to be more research to improve these means or otherwise make them more cost effective.

Even greater savings in energy can be achieved if greater use in agriculture could be made of biological nitrogen fixing. Generally known is the role of rhizobia, the nitrogen fixing bacteria that adhere to nodules on the roots of legumes. These are anaerobic procaryotes, which derive their energy from carbohydrates in the plant, reducing net photosynthesis. Consequently, the yield of legumes tends to be lower than that of competitive crops. However, it is estimated that legumes get only about 25% of their nitrogen from rhizobia. It is evident that there are other nitrogen-fixing microorganisms in the soil that are photosynthetic, deriving their energy from the sun and not from the host plant. Some are beginning to be identified. There are trees, such as the ginkgo, and ferns that have nitrogen-fixing microorganisms associated with them. Anabaena, the blue-green algae associated with the azolla fern, is photosynthetic and has been a nitrogen source in Chinese agriculture.

Brady notes that in the last decade valuable research has added important knowledge on the biological nitrogen fixation process. The research agenda on biological nitrogen fixation has considerable possibilities for world food and agriculture. The most commonly discussed possibility is the breeding of varieties of rhizobia that will adhere to the roots of nonleguminous plants, particularly the cereal grains. It is also argued that there is considerable possibility for improving the efficiency of the nitrogen fixation rate of rhizobia. Another possibility is to find ways of increasing the output of free-living nitrogen-fixing microorganisms. There is, in addition, the potential of breeding nitrogen-fixing capability into plant cultigens. Alternatively, one could attach photosynthetic nitrogen fixers to the leaves of plants, as is the case with some trees. Finally, with genetic implant, the efficiency of free-living, photo-synthetic microorganisms such as the blue-green algae can be introduced more extensively in agriculture, particularly in the humid tropics.

Plants, like humans, are subject to a phenomenon called stress. Plant stress can result from a variety of factors. Problems of soils and lack of nutrients can be a cause of stress. Extremes of temperature and excess or deficiency of water are other causes. Stress can either kill or stunt the plant.

Water performs many functions for the plant. It is a source of hydrogen in photosynthesis. Water helps in transporting nutrients in the plant from where they are created to where they

are needed.   In transpiration, evaporation of water plays a role
in regulating the plant's temperature.

WATER AND AGRICULTURE
     Water has been a plant requirement and humans have been
artificially supplying it since the earliest agricultural
systems.   Agricultural systems were constructed around the
availability of water, whether from a rainy season or an annual
flooding of a river.   In some quite arid climates, landforms were
modified to direct and concentrate the flow of water to a patch
of land that was being cultivated.   In drier seasons along
rivers, water was brought up and fed to the soils by a variety of
animal-powered waterwheels in some of the earliest forms of
irrigation.   In time, humans would learn to contain or otherwise
dam up water and save it for release at more critical periods
during the growing season.   If the topography was right, the
storage areas could release the water and allow it to flow by
gravity along previously dug canals.   Where the lay of the land
was inadequate for gravity flow, human and animal power was
necessary.   Digging wells or using naturally flowing artesian
wells was yet another means of obtaining water and directing it
to support plant life.
     The land creating and land extending (by adding to the
growing season) accelerated after World War II.   Large-scale (and
small-scale) dam projects provided the containment for water and
the electric power for pumping, if that was necessary.   Rural
electrification and diesel pumps provided power for pumping
water, whether the source be a river or ground water.   New
varieties of plants bred for shorter growing seasons facilitated
multiple crop seasons and required the use of irrigation in areas
where rainfall had been sufficient for the traditional crop
season.
     Currently, the estimates of the land being irrigated range
from 200 million to 250 million hectares, or about one-sixth
of the world's agricultural land.   This is about half the
potentially irrigable land.   However, this one-sixth of the land
produces approximately one-half of the world's harvested output
(Tekinel 1979, 464-65; Slater 1981a, 100).
     The total quantity of water on earth is not a limiting
factor to human agriculture, industry, and human habitation in
general.   Ambroggi (1980, 101) says, "global reserves of fresh
water add up to more than 37 million cubic kilometers."   This
"fresh" water is about 3% of the total water resources of the
earth.   Ninety-seven percent of the earth's water is in the
oceans.   About 66-75% of the remainder is contained in the ground
(Ambroggi 1980, 101; Tekinel 1979, 457).   The 97% of the water in
the oceans is available for human use only for transportation via
ships, as a food source (i.e., fish--about 2% of world food
production), and as a minor source of minerals.   With current
technology, its direct use as water is limited to the very costly
process of desalinization.   However, indirectly the oceans are
critical as a source of water taken up through evaporation and,
dispersed as rains.

The 2 main sources of water for human use are from rain and its subsequent runoff in streams, rivers and lakes, and from ground water, which can be fossil water (and therefore not sustainable if drawn up) or part of the current water cycle. Even aquifers that are currently active can be drawn down at a faster rate than they are replenished.

The oceans are the source for 430,000 of the 500,000 cubic kilometers of water evaporated each year. Of these, 390,000 cubic meters are returned directly to the ocean, while the land surface receives 110,000 cubic meters for a net gain of 40,000 cubic meters of fresh water. Of this amount, Ambroggi estimates that 14,000 cubic meters are available for human use. The water cycle is complete when the net water gain is returned to the ocean through surface runoff. Subtraction of the 5,000 flowing in areas not currently suited to human habitation leaves a water budget potential of 9,000 cubic kilometers. Ambroggi estimates that this 9,000 cubic kilometers could support a world population of 20 billion to 25 billion people. The limiting factor in water for development, then, is its distribution "from place to place and from season to season". Roger Revelle (1973, 92) estimated that "a billion more feet per year or less than 4% of the total river flow is used to irrigate 310 million acres of land, or about 1% of the land area of the earth".

Even in the humid tropics there are periods of the year when water stress limits output (Brady 1980, 19). Too much water can be as much a cause of stress as too little. In fact, a significant cause of waterlogged soils is irrigation. In addition to raising the water table, irrigation can bring transformations of the soil adverse to sustained agriculture. Old and new techniques are continually being explored to either remove excess water or to bring irrigation to additional lands. These range from simple processes of digging tube wells and installing small pumps to using satellites for mapping and computer programs in developing models to facilitate artificial recharging of aquifers. Of course, both can be part of the same project, as water stored in an aquifer can be brought up by individual farmers through tube walls. Improving drainage and lowering water tables by pumping the water out onto new lands are but two of many strategies for attacking the problem of excess water.

One of the ways to handle the problem of scarcity of water is to use it more efficiently. The work that the Israelis pioneered in drip irrigation is widely known for its controlled ability to derive maximum benefit from a small amount of water. One of the most promising ways of conserving on groundwater (also giving the soil some protection against erosion) is no-tillage or minimum-tillage agriculture. The no-tillage cropping system is described by one group of authors (Phillips et al. 1980, 1109) as being "a combination of ancient and modern agricultural practices." It is ironic that these dry-farming practices that are now considered to be useful are similar to traditional agricultural practices in the tropics that were so severely criticized by colonial agricultural officers.

No-tillage (or reduced or minimum-tillage) involves a reduction or elimination of plowing in the traditional sense. To the extent that plowing is seen as a "labor-saving invention for killing weeds," then the substitution of selective herbicides for some plowing can help to preserve the structure of the soil and the water in it (Giere et al. 1980, 15). There are also some water losses through evapotranspiration of the vegetation left on the field.

The central question of this chapter is not whether we can use chemical nutrients and herbicides and pesticides to increase world output of foodstuffs. As figures on the growth of food production in subsequent chapters demonstrate, we have done so successfully and are likely to continue to do so. Nor is it a question of whether we are approaching limits to soil fertility or global water supplies. A basic issue is whether our current practices are not only expandable but, more important, sustainable. Equally critical is whether or not cultivation methods, water use, seeds, and chemicals can be improved, invented, and adapted so as to be usable by poor farmers around the globe. It is not whether with modern science and technology we can collectively wrench out a sufficient global food supply, but whether poor farmers (particularly in Africa) can gain access to the credit, the knowledge, and the instruments necessary to increase their output and income. For many parts of Asia and Latin America, improved technical packages are available; still, these things are not always available for use by small farmers. For most of Africa and some other areas of the tropics, substantially improved agricultural crops and practices are not really in existence. In some instances, then, the technical means are available, but it is still necessary for scientific and technical research to redesign them so that they are more readily accessible by poor farmers. In other instances, attacking the hunger problem requires more fundamental advances in agricultural research for the food crops and climates of regions such as tropical Africa.

Too often in the past decade the issue of hunger and food supply has been framed in global terms. Also, there have been tendencies to project population growth and food supply expansion into some often distant future where catastrophe will result. The problem is not global, as food supply growth has been unprecedented and continues to be expanding, though at reduced rates. Global food sufficiency does not help the farmers and countries in Africa, where per capita (sometimes absolute) food production has been falling; surplus food does not help those who do not have the income to buy it; nor do new technologies necessarily benefit those without credit to gain access to them; and it is not the global limits to growth that are troubling, but it is the particular environmental limits that many of the world's farmers are operating within.

As this is being written, some of these regional limits are so severe, such as drought in many parts of Africa, that the technologically sophisticated farmers are suffering with their poorer neighbors.   In other years and other instances, the environmental limits are not those set by nature but those set by limited access to technology.   We live in a world where some large regions are worried about agricultural surpluses and prices insufficient to sustain future output, while in equally large regions or areas hunger and its concomitants of malnutrition, disease, and death stalk the land.   It is less a case of whether we can produce than of who can produce so as to either feed themselves or gain the income to obtain sustenance from others.

CHAPTER 9
Food Supply and Nutrition

The question of a food crisis is one fraught with emotion and one that has been the subject of outrageous overstatements in the past. First, what is the food crisis? The specter of famine of the kind that has been the periodic lot of much of mankind throughout history is not in the offing on a global scale. However, some areas, such as eastern and southern Africa, are facing severe drought and famine. In fact, if one takes the world's food production, following the procedures of the Food and Agriculture Organization (FAO), subtracts a generous estimate for loss through pests, rotting, etc., converts it into kilocalorie equivalents, and divides this result by the world's population, one finds that current production provides about twice the amount considered necessary for minimum daily caloric needs. Because part of this food output is fed to livestock, that which is available for human consumption is greatly reduced. However, this reduced quantity is not a function of limits to the earth's food-producing capacity but is the result of the distribution of income and the pattern of demand for food that results from it.

Contemporary food problems also are not the result of any failure to make progress on the issue since World War II. In fact, one can only describe the changes in food agriculture over the last 3.5 decades in superlatives. They are unprecedented in human history. From the mid-1950s to the mid-1970s, world food output per capita increased by over 20% (Wittwer [1977] 1978, 487). In the third world, in roughly this time period (1961-1977), major food crop production increased at 2.6 percent per year (IFPRI Report 1982, 14; Lewis 1983, 116-17). With population growth, the per capita growth was smaller but still positive. There were advances in other food crops as well. This per capita increase in output was occurring at a time when the world's population was increasing by almost 50%, thereby requiring nearly a doubling of total output to achieve this per capita increase.

POPULATION AND WORLD FOOD SUPPLY
Increased and stabilized food production was a factor in this rapid expansion in population. Population growth was the

result of a fall in the death rates; the primary causes of this
fall were the containment (and occasional elimination) of many
infectious diseases and the improvement of the food supply. The
2 are closely related because ill-nourished people are far more
susceptible to disease than are the adequately fed. Roughly
about 50 million people died in the world in 1950, and about 50
million (probably a little fewer) died in 1975. Had the 1950
death rates prevailed in 1975, an extra 25 million people would
have died. The number of deaths this year will probably still be
pretty close to that number (Mauldin 1977, 395; Eberstadt 1980,
40).

Population was increasing even though birthrates were
falling in most parts of the world. In third world countries,
the decline in the rate of live births was 3 times the fastest
rate that births had fallen in Western countries (Eberstadt
1980, 42). However, population growth rates increased because
"Third World mortality decreased five times faster than it
did in the developed countries when they were at a similar
stage of development" (Loup 1982, 4). In third world countries
the annual real per capita economic growth also was averaging
about 3.4% per year, and literacy and other indices of development experienced rapid increases (Morawetz 1977, 10). Using
Paul Bairock's estimates, Jacques Loup compares a 0.8% Gross
Domestic Product (GDP) per capita growth from 1900 to 1952-1954
to a 3.0% GDP per capita per year increase from 1950-1980. "On
the average," writes Loup (1982, 19), "between 1950 and 1980 the
gross domestic food produce per capita doubled." These changes
were unanticipated in the magnitude in which they occurred, even
by the most optimistic economists and politicians.

These varying successes of food production and other forms
of development were creating the problem of population growth
and its attendant difficulties of too rapid urbanization,
occasional social disruptions, and pollution and other forms
of environmental degradation. Birthrates were falling but,
unfortunately for population control, death rates were falling
faster. Loup reports the decline in the birthrate was 1 point
from 1960 to 1965, 1.7 points from 1965 to 1970, and 2.9 points
from 1970 to 1974. Even here, by the late 1960s it should have
been obvious that birthrate declines worldwide were accelerating
while death rate declines were slowing. By the mid-1970s birthrates were falling faster than death rates, bringing about a
fall in the rate of population growth. This fall in the rate
of growth would still bring a population increase, but most
population projections for the year 2000 had to be revised
downward by a couple of billion. Population growth has its own
momentum. Past high growth rates mean a young population with a
large portion of the population yet to enter the childbearing
years. Continuation of declining birthrates and continued
increases in life expectancy will in time mean that death rates
will rise further helping to close the gap between birth and
death rates (Coale 1983, 828-32).

Ironically, it was toward the end of this period, from 1950

to 1970, that a pessimistic dissent on development emerged. Granted, there were some causes for concern. People don't eat averages, and development was unevenly distributed about the globe. Within some countries, moreover, economic inequality had increased. It was even argued that in many instances the poor were actually worse off as a result of economic growth. Valid though many of these concerns were, there is still no denying the significance of increased per capita food supply and lengthened life expectancy. When dealing with money income, a poor majority with falling income can be offset by a minority with large and rapidly growing income that creates misleading rising national averages. It is difficult to imagine, however, the affluent devouring all of that increase in food supply, even allowing for a substantial conversion of grain to meat. Furthermore, increases in life expectancies at birth (1950–1955 to 1970–1975) such as 47.5 years to 63.3 years in East Asia, 37.5 years to 46.5 years in Africa, 52.0 years to 61.2 years in Latin America, 39.2 years to 49.3 years in South Asia, and from 42.5 years to 53.2 years in third world countries overall can in no way be explained by increases only for a few (World Bank 1980, 11). Improvements in life expectancy were higher both relatively and absolutely in third world countries than in industrialized countries.

Although the late 1960s and early 1970s brought an outburst of catastrophic predictions, the massive disasters failed to materialize. However, some of the more thoughtful criticisms of development contained some merit. While the world didn't go to hell in the 1970s, the improvements of the prior two decades slowed appreciably. Economic growth also encountered increased difficulties. Improvements were paid for by increasing levels of the poor countries' indebtedness, which rose dramatically after the 1973 oil price climbed. Competent scientists, moreover, began warning about the dangers of man-induced species annihilation, climatic change, and environmental destruction. The demographic transition (birthrates falling more rapidly than death rates) also occurred during the 1970s, but this was primarily the result of changes already in progress.

An outburst of concern about mass starvation came at the same time as the visions of resource exhaustion. From an economist's view the evidence against both prophecies is similar. Real resource costs have been falling. The same is true for world food costs. There has been a 100-year decline in the real price of food and at least one author, Clifford Lewis (1983, 106–7) expects it to continue. He notes that "despite the weakness of data about global food trends, it is clear that global grain production has grown at a generally increasing rate since the early decades of the nineteenth century and has consistently outstripped increases in global population." Lewis shows the real price of wheat per bushel (in 1967 dollars) falling from $3.09 in 1890 to $2.10 in 1960, to $1.10 in 1982. There have been "temporary disruptions" in this downward trend, but they have been "when peace has not prevailed."

When one breaks down the averages for food production the

1970s into countries and regions, the picture becomes more clouded.   On the favorable side, countries such as India and China have survived bad weather to continue to expand food production.   In agriculture there is no getting away from the weather, but modern agriculture and irrigation provide a greater hedge against adversity than most people realize.

The unfavorable side of the 1970s gives some indication of where some of our difficulties lie.   Population continues to grow rapidly in tropical Africa.   There is some debate among demographers of Africa; the evidence points toward high birth and death rates, neither of which have been falling in recent years.   Food supply has been increasing less rapidly than population in most countries of tropical Africa, and there are several countries where food production has fallen in absolute terms.   Wars, political instability, droughts, and refugees have exacerbated the food problems, but these cannot be used as the sole explanation for them.   Policy errors, particularly in agricultural pricing and urban food subsidies, are seen by many economists as the major cause of food production decline in Africa.   Along with political and natural calamities, they contribute to the current difficulties in Africa.   If per capita food consumption has not fallen in all these countries of tropical Africa, it is because they, like other food-deficient countries, have made up the difference with imports.   These imports have tended to further complicate the poor countries' problems of international indebtedness.   The irony is that African countries generally performed well, improving over colonial performance in both agricultural and manufacturing output during the 1960s, the first decade of independence for most of them.

There is concern, even among those who pioneered the green revolution and other such scientific research, that the pace of the past decades cannot be sustained (Borlaug 1981, 112-13). Though not pessimistic, Borlaug believes that we are close to the genetic limit on some of our major crops (wheat and rice) and that our "breeding program appears to have reached a genetic yield plateau" in wheat.   The diminished prospects of some people for food improvements in the 1980s are in part a function of the factors that accounted for a large part of the increases of the last 3 decades.   Important in these past advances were (1) new seeds, particularly the "miracle" wheats and rices; (2) a period of exceptionally favorable weather for agriculture in the main growing areas of the world in the late 1950s and early 1960s; and (3) the spread of energy-intensive agricultural techniques in a period of cheap energy.   Clearly, the same constellation of events cannot be assumed for the coming years.   Cheap energy for agriculture is obviously not in the offing.   Drought plagued many lands in the 1970s, and there are climatologists who say that the good weather of the 1950s and 1960s was atypical.

The future is further complicated by the level of indebtedness that third world countries have incurred in the last few years in an attempt to maintain their previous levels of energy imports.   It is questionable whether they will have the capacity

to service and expand this indebtedness at a level to allow them
to sustain current food output, let alone expand it. However,
there are special provisions that can allow for critical food
imports, even where indebtedness and payment problems lead to the
virtual disappearance of other imports. Food import dependency
is becoming a common phenomenon throughout the world as countries
increasingly rely on the United States, Canada, Australia, and
Argentina to provide grain seeds. Many argue that the provision
of food aid, coupled with government policies that keep the price
of food down, has created major disincentives that have held back
agricultural production in the third world.

        This food interdependency between producer countries and
importing nations has created a dangerous dependency for both
buyer and seller in a debt-burdened international trade. Agriculture in producer countries has become dependent on sales
in the global markets to sustain the economic and technical
structure of its productive processes. Many importing countries
have become dependent on borrowing to finance their food imports.
With many countries borrowing to their limit and beyond, it
is possible to have worsening food deficits and consequent
malnutrition in poor countries while exporting countries are
experiencing food surpluses that threaten prices and undermine
the financial solvency of the farmers and agricultural community.
Food interdependency might be more accurately described as mutual
dependency. This mutual dependency is the result of a growth in
the volume of world trade in food of 5–6% per year since 1950.
The increased production and trade in food result in a compound
annual growth rate in world per capita consumption of .85 for the
years 1950–1981. This rate was 0.95 in 1950–1972, falling almost
by half to 0.50 for the period 1967–1980. Aggregate production
of consumption grew 2.7% per year over this period (O'Brien
1982, 2–3, 12). In the few instances for which we have data
before this century, it is rare to find countries, even those
experiencing astounding agricultural transformation, averaging
more than 0..1% per year per capita food production growth over
the course of a decade.

AGRICULTURAL IMPROVEMENTS AND THE GROWTH OF OUTPUT

        Some prognoses for the year 2000 portend large-scale food
deficits in the developing world. Many who make these projections are by no means catastrophists. The possibilities
inherent in the scientific and technological research discussed
in these chapters are more than sufficient to close this gap. As
with research of any kind, however, there is no clear certainty
that discovery will keep pace with need, nor that possibilities
achieved through research will be fully realized in the field
of production. Even some of the more optimistic projections
for the timetable of major breakthroughs in basic agricultural
research make it less than certain that our goals for the
year 2000 will be achieved. Still, however spectacular or
unspectacular advances are likely to be, there will be some
continuous improvements in agriculture.

Duvick (1983, 578) writes, "In no crop is the curve for genetic yield increase starting to level out. In all crops, expression of gains in genetic yield potential does not necessarily require changed fertilizers." Though the new crops do not require more fertilizer, they are generally more responsive to agricultural inputs. He says modern agronomic practices have greatly enhanced the yield potential of the newer cultivars, "whereas yield potential of the older cultivars is enhanced only slightly or sometimes even decreased by such modern cultural practicies as high plant density and high nitrogen fertilizer applications."

Duvick further found that empirical testing of the newer cultivars has made their yield more stable in response to environmental stress. The new cultivars are tougher, he says. To whatever extent we are successful over the coming decades, it is inconceivable that a large measure of this progress will occur without scientific and technological research.

A regional breakdown of the figures on international food trade reflects changing patterns. Fifty years ago every region but Western Europe was a net grain exporter. For the last decade, and in growing magnitudes, every region but North America, Australia, and New Zealand has become net grain importers (Journal of the Society for International Development 1982, 5). It has been the massive export of wheat from the United States, Canada, and Australia, and to some extent, Argentina, that has supplied the growth in output and consumption in food.

Even regional aggregates do not tell the entire story. The same change in import-export patterns will have different meanings for different countries. An increase in food imports into Singapore and Hong Kong reflects population growth and successful economic performance. Nobody expects these urban enclaves to be self-sufficient in food any more than one would expect New York or Paris to be. In other instances, rising imports can also reflect economic growth as urban consumers diversify their diets. Importing coarse grains to feed animals is not necessarily indicative of domestic agricultural failure.

If some countries and regions are doing far worse than the average, then obviously some are doing better. Beyond the performance of energy-intensive agriculture of developed regions of North America and Australia that have been pouring on the fertilizer to increase the yield, there are a number of countries that have done remarkably well in either increasing their domestic food supply or in transforming their economies to enable them to buy food from abroad. These are essentially the same countries that have dramatically lowered their birthrates and extended their life expectancy. The performance of countries such as Korea, Taiwan, and Singapore is so unprecedented that it is difficult to describe it without use of superlatives. Had anyone forecast these countries' rates of transformation 30 years ago, they would have been deemed utopian.

The aggregate food output figures can conceal other changes.

Increasingly large amounts of grain are diverted to feed animals, which in turn are used to feed people. The ratio of grain to meat output can vary anywhere from 2 or 3 to 1, to 6 or 7 to 1, depending on whether poultry or cattle is being raised. Thus, someone from the United States will eat 4 to 5 times as much grain as the typical Indian consumer. However, the person from India will eat more grain as grain, while the American will eat more than 80% of his in the form of grain-fed animals. It is likely that neither of these circumstances is nutritionally optimal.

There is another adverse change that can be masked by aggregate statistics. In many parts of the world, population growth and increased yields per acre from the high-yielding grains have combined to induce farmers to switch lands from pulses or other vegetable crops to grains. This expanded output might meet the caloric needs of a population while unbalancing the diet by reducing the intake of vital foodstuffs. This does not mean that a grain such as wheat is an inferior food. Far from it; it is frequently used as a major food for crisis intervention. It is simply used that wheat alone, like any other foodstuff, is not a complete foodstuff. Whatever else is said about unbalancing diets, however, the net effect of the changes in food availability has almost certainly been positive. While the extension of life expectancies almost everywhere has many causes--public health, innoculations, and so on--it is highly likely that the expansion and regularization of food supply has made an important contribution.

Wheat illustrates the dramatic changes that have taken place in food production. Wheat and rice are the 2 most important foodstuffs in the world and are also the ones in which the most dramatic gains have been made. "Since World War II," write the authors of Wheat in the Third World (Hanson et al. 1982, 15), "more progress has been made in raising wheat yields of developing countries than occurred in the preceding 8,000 to 10,000 years following domestication of the crop. In 1950 the average wheat yield in developing countries was about 700 kg/ha, but by 1979 it had risen to 1450 kg/ha." Even this later yield is far below the world average of 1,900 kilograms per hectare and the developed country average of 2,270 kilograms per hectare. The authors also give other yield figures up to the world record of 14,000 kilograms per hectare and the theoretical maximum, 20,000 kilograms per hectare. Clearly we are not near the limits of wheat production, even on land currently in production, using currently available technology.

Through much of human history, increases in agricultural production have come from expansions of the area under culti- vation (Evans 1980, 388). As a result of the dynamic interaction between humans and plants, there was selective breeding that improved yields. These increases in yields were erratic, with occasional retrogressions, and are most evident when viewed over long periods of time. Likely wheat yields in early times down through the European Middle Ages were on the order of 0.5-0.75

tons per hectare. At various times in this period, favorable growing conditions may have given yields as high as 2-3.6 tons per hectare. These compare with more modern yields that range from 10 to 14.5. Over the past few centuries, yields increased fairly steadily but slowly. The curve of growth started to turn up in the nineteenth century as a result of the developments in agricultural science. Viewed over this long historic period, gains in animal production were even more dramatic. Brown (1974, 23) writes, "The first domesticated cow probably yielded no more than 600 pounds of milk per year--barely enough to support a calf to the point where it would forage for itself. In India, milk production remains at about that level today. By contrast, in 1973 the average milk cow in the United States yielded over 10,000 pounds annually." In roughly the same timespan, egg production rose from about 15 eggs per chicken per year to well over 200.

Most of us have heard the story of the plant breeding activities at CIMMYT, the International Maize and Wheat Improvement Center in Mexico City. Norman Borlaug won the Nobel Peace Prize for his work in breeding the dwarf high-yield varieties (known as HYVs) of wheat. The miracle wheats and rices have been the object of considerable hysterical reaction by many who see a global catastrophe resulting from their use. Wheat in the Third World provides a calm, factual rebuttal to these charges. It shows how the plants were bred for disease resistance and notes that "it is significant that stem rust epidemics have not recurred in North America and Mexico in more than a quarter century" (Hanson et al. 1982, 24). The authors draw on Grant Scobie's (1979) World Bank study of high-yielding varieties of wheat and rice. They found that, though there were increased regional variations in income, overall the new varieties brought improved nutrition, increased demand for agricultural labor, and greater benefits proportionally for lower-income groups. The greatest beneficiaries of all were the low-income consumers.

The dwarfing and semidwarfing varieties were developed by the "conventional breeding procedures" of the 1940s and 1950s. These generally "required 10 to 12 years to produce a new wheat variety," though in Mexico they did take a few gambles, accelerated the process, and, luckily, won (Hanson et al. 1982, 31). Today, with gene splicing and cloning, these techniques seem positively ancient history, though they are still used and produce benefits. Now there is talk of rebuilding the plant, improving photosynthetic efficiency, creating nitrogen-fixing bacteria (or even better, with a photosynthetic organism) for cereal grains and a host of other breathtaking possibilities. While many of these make good news items, most of the sensational breakthroughs are 15 to 50 years away. Even assuming that all these things happen, there are and will be a lot of hungry mouths to feed until the commencement of the new utopia. Nevertheless, the same advances in the sciences that give rise to fundamental research in plant genetics are also providing continuous advances in practical understanding, from how to use fertilizer more efficiently to new forms of food preservation and to the continuation of yield-improving plant breeds.

Lewis (1982, 117) gives 3 basic reasons for believing in the "technical capacity to produce the supply" of food. They are that (1) land resources are available; (2) there is room for yield improvement in land-scarce country; and (3) continued genetic improvement can be expected. Further, he argues that the financial constraints are not as great as has been argued. Much of the current high cost of food imports is for foods for higher income groups. In a crisis, these could be cut to save exchange for the critical grain and other basic food imports. The international food safety net guarantees food at concessional prices and drawing rights on the International Monetary Fund to finance it, which "are not subject to IMF conditions" (Lewis, 110-11). Both past performance and current evidence support Lewis' contention that except for disruptions caused by war, other political disturbances, or extreme variation in climate, there is a reasonable expectation that growth in food supply and the ability to purchase will continue to outpace growth in population.

HUMAN NUTRITION

Food provides the ingredients for the building and functioning of animal life. Humans, like other life forms, take in energy, use it to carry out activities, and store the rest. That which is "used" is thereby released in a degraded or less usable form. Plants take the energy directly from the light of the sun and, through photosynthesis, convert it to carbohydrates. Respiration is the reverse of photosynthesis. Photosynthesis takes carbon dioxide, water, and energy; manufactures glucose; and gives off oxygen. Respiration takes in oxygen; breaks down hydrocarbons, giving off carbon dioxide; and releases energy for use by the life form. Plants also use the process of respiration for energy needs, "burning" the carbohydrates that they have created. There is thus a distinction between gross photosynthesis--the total energy taken in and used to create glucose--and net photosynthesis--the energy, carbon, hydrogen, oxygen, and other nutrients that constitute the plant structure itself. These stored nutrients, the result of net photosynthesis, form the base of the food chain upon which virtually all life on earth is based.

A plant not only takes in energy but also has the ability to use elements in the air and soil to synthesize the amino acids necessary for its life structure. Animals, including humans, have a more limited ability to manufacture their needs, and so must derive both energy and vital compounds. Carbohydrates and sugar provide energy but are not the only source of it. All animal cells contain protein.

There are 20 amino acids used in the biosynthesis of proteins. The amino acids are all characterized by the presence of a carboxyl (COOH) group with acidic properties and an amino

(NH_2) group with basic properties, attached to the same carbon atoms; the rest of the molecules vary with the particular amino acid. The structure of an amino acid may be represented by the formula

$$R - \overset{\displaystyle H}{\underset{\displaystyle NH_2}{\overset{|}{\underset{|}{C}}}} - COOH$$

where R represents the remainder of the molecule (Davidson et al. 1979, 34).

The amino acids are bound together in chains, by bonds called a peptide linkage, to create a molecule of protein. Each species of protein has its own unique set of linkages and combinations of sequences of amino acids. Though the human body cannot manufacture amino acids de novo, it has a limited ability to synthesize some from other amino acids in a process called transamination. Those that the body cannot manufacture are called the essential amino acids. These eight are isoleucine, leucine, lysine, methionine, phenylalanine, threonine, trytophan, and valine. Histine is also needed by children for growth.

The growing organism needs protein to build the cells and structure of the body, just as it needs energy to carry out its functions. Both the growing and the mature organism are continuously experiencing wear as the body functions. Thus, proteins are needed for repair and replacement of cells. The body itself releases protein that is derived from discarded cells. The rest has to come from food. Just as the growth of plants is limited by the least available amino acid, so also is the body's growth, repair, and replacement limited by the least available amino acid. The intake and expulsion of amino acids is called the nitrogen balance.

The very concept of food and its nutritional value is relative to the organism that uses it. According to Bressani and Elias (1973, 254), "A highly nutritious food could be defined as one which is utilized by the animal organisms with a minimum of waste, alone or with other foods, providing adequate amounts of the nutrients needed to meet the demands for the physiological functions of the organism." Similarly, the biological or nutritional value of a protein source is dependent upon the correct balance of the amino acids. Thus, limiting amino acids means others are wasted and cannot be used to build protein.

Most of the plant material we eat is in fact a stored food/energy source for the plant itself or for its offspring. When plants synthesize carbohydrates they use some to build the structure of the plant and they store some. Of the stored energy, some is used during periods when photosynthesis is not active and others are used as a food energy source for the offspring of the plant. "Man and his ancestors have been remarkably successful in seeking out and later cultivating

various seeds, fruits, and roots which contain concentrated
supplies of carbohydrates," write Davidson et al. (1979, 26).
The authors contrast human food sources with that of ruminants
(sheep, for example) that have to spend most of their life
chewing grass, which is 80% water. Humans, by cultivating
high-energy foods, can store the product, use it when needed, and
have time left for other things, such as building civilizations.
Humans have adapted plants to meet their needs in the continuing
process of domestication. In general, the changes have facilitated the growth and reproduction (particularly for adaption to
different environments) and enlarged the concentrated energy
source (such as seeds) relative to the rest of the plant. There
have also been the selection and creation of varieties with
different taste characteristics. However, until the last few
decades there was essentially no thought of breeding a plant
food source that more completely met human nutritional needs.
What is balanced nutrition for the plant is not necessarily
balanced nutrition for the organism that feeds off the plant.
Consequently, the definition of nutritious food recognizes
complementary foodstuffs are needed for balanced efficient
utilization.

There are other nutrients that the human body needs in
addition to the essential amino acids. Humans, almost by
definition, have found ways of balancing their diets (i.e., we
have survived to the present). That does not mean that we have
always done it efficiently or well. After all, malnutrition has
been and remains a widespread condition in human societies. In
developed countries, the range of foodstuffs available for a
balanced diet is so great that for most, malnutrition is a
function of unwise choice rather than of lack of opportunity. In
poorer societies, balancing a diet frequently consists of
combining a grain (in which lysine is the limiting amino acid)
with a pulse (in which methionine is the limiting amino acid).
In a food regimen such as this, depending on the culture,
location, income level, etc., the diet could be filled out with a
variety (each in small quantities) of other vegetables such as
spices, soups, or side dishes. If not prohibited by belief
systems, small amounts of meat are sometimes available. When
population growth leads to a concentration on one foodstuff,
such as a dwarf grain with a high energy yield per hectare,
nutritional deficiencies can result.

Humans can regulate the mix of foodstuffs they consume in an
attempt to achieve optimum growth. In theory at least, one could
regulate human food intake in the same way that we attempt to
regulate plant nutrition. If the deficiency is in calories, then
we add calories to the diet. If the limiting factor is an amino
acid, then we add that amino acid. Whatever the deficiency, a
vitamin or a mineral, we add to the diet. Unfortunately, the
reality is much more complex, particularly when one is trying to
help improve the diets of those whose income is low.

Diversifying the food types consumed is the way most people
balance their diet. For those for whom the quality, quantity,

and diversity of the diet is limited, there are 3 techniques for improving the foods that are consumed. They are fortification, restoration, and enrichment. Fortification is defined "as the addition of specific nutrients in higher amounts than that found in natural foods." Restoration is defined "as the addition of a nutrient to a particular food to restore its original chemical composition." Enrichment involves the addition of nutrient(s) to a specific food to utilize it as a carrier for this nutrient(s) (Bressani and Elias 1973, 259-65).

With fortification one could, for example, add a synthetic amino acid, such as lysine, to wheat. By increasing the limiting factor, this in theory increases the availability and efficiency in using all the amino acids. An example of restoration would be returning to bread nutrients that are removed from the grain in milling. An example of enrichment is the addition of iodine to salt where it has been used as a cure for goiter. This is an instance where one could reasonably argue that one of these techniques has worked and at a very low cost. All of these techniques have been and are being used in developed countries. When these techniques were introduced, a variety of nutritional disorders in these countries were virtually eliminated. However, a vast number of other changes were taking place (the U.S. economy moved from depression to war, to postwar prosperity) and there is no verifiable evidence that these specific interventions were either the necessary or the sufficient cause of the nutritional improvement. Some interventions have clearly worked, such as iodine for the elimination of goiter, while for others there is insufficient evidence for their efficacy or lack of it.

Speaking generally about the more complex criteria for improving human diet, J. H. Hulse (1982, 1292) argues: "Nutritional studies with human subjects are expensive and time-consuming; humanitarian and ethical considerations restrict the range of variables that may be studied, the techniques of assessment that can be used, and the period of time over which any experiment may be continued." Concluding an article summarizing the historical practices of food fortification, restoration, and enrichment, L. J. Teply (1973, 247) concludes that while "food fortification is not a panacea, this should not detract attention from what appears to be vast possibilities for wider application of this approach."

MALNUTRITION

No discussion of nutrition is complete without mention of malnutrition and its consequences for the most vulnerable sufferers from it, children. Two types of nutritional deprivation in children are called marasmus and kwashiorkor. Traditionally marasmus was seen as a caloric deficiency, while Kwashiorkor was defined as a protein deficiency. More recent theories see them both as resulting from essentially the same nutritional deficiencies with the growth retardation of marasmus representing a more effective way for the body to respond to dietary stress (Caliendo 1979, 7).

Ideas about nutrition in poor countries have passed through various phases in the past decades, at different times stressing protein deficiencies, caloric deficiencies, or deficiencies in a specific vitamin or mineral. Currently much of the literature speaks about protein-caloric malnutrition, or PCM.

Marasmus and Kwashiorkor are seen by many as just extreme manifestations of what is a much more widespread condition of malnutrition. Not all slower growth rates imply a condition of malnutrition. Mellor and Johnston (1984, 546-47) write, "The adjustment of children to a slower growth trajectory and a smaller attained size may or may not have serious consequences, depending mainly on the severity and timing of food deprivation and on sanitation and related factors." Some argue that people can be small but healthy (Seckler 1980, 219-27).

Kwashiorkor is a term used by Dr. Cicely Williams, who in the 1930s described the condition. A word in the Ga language in West Africa, it literally means the affliction of the child who is displaced from the breast by a succeeding child (Caliendo 1979, 13). The early weaning of the child in a food-deficient environment can quickly lead to protein deficiency. The starchy gruel of various kinds used to wean children lacks sufficient amino acids for continued growth. If the body does not have the essential amino acids needed to manufacture protein, it not only cannot support growth, it can no longer maintain itself by replacing the cells and tissues that are worn out.

Without sufficient protein the body must ration it to provide for the vital needs. When ingested food is inadequate, the body will start to digest and utilize its own less vital protein tissues in a process called the "metabolism of wasting" (Whitney and Hamilton 1981, 138). Hair growth ceases. Skin is no longer manufactured, therefore sores and other lesions do not heal. This reduces the body's defenses against invasion by disease. "Many of the antibodies are also degraded so that their amino acids may be used as building blocks for heart, lung and brain tissue" (Whitney and Hamilton 1981, 139). With fewer antibodies, the child's resistance to disease is further reduced. One disease frequently contracted is dysentery, which causes an additional loss from whatever nutrients the child may be obtaining. Any parasite (by definition) feeds off of the host organism (in this instance, the child), increasing the nutritional deprivation (Ash et al. 1984, 17-20). Thus there begins a worsening cycle of malnutrition and disease. In some instances the situation is further complicated by the mother trying to stop the diarrhea by withholding water or nutriments.

Edema (also spelled aedema), the accumulation of fluids resulting from hormonal imbalance, gives the kwashiorkor child a puffy look that partially masks the emaciated condition. Marasmus gives a wizened, emaciated look that is sometimes described as making the child look old. The lowered metabolism, a response to the reduced caloric intake, makes the child inactive as a way of conserving energy. The body temperature is below normal and the lack of fat beneath the skin reduces the child's protection against heat loss in cold weather. Like the kwashiorkor child, the child suffering from marasmus is susceptible to disease.

Most of the growth of the brain occurs in the first 2 years of human life. Nutritional deficiencies at that time can retard the development of the brain and result in permanent disability in learning. The child whose energy and activity level is low is likely to get less of the behavioral responses from adults necessary for cognitive and social development. Even in adults malnutrition can lead to a phenomenon called the "listless poor" (Scrimshaw 1984, 21, 24). Similarly, this leads to a negative response: others blame the condition on laziness and not on other causal factors. The consequences of severe malnutrition are either cycles of disease and worsened nutrition, leading to death, or cycles of ill health and inactivity, making it more difficult to obtain the quality and quantity of nutrition necessary to break the cycle. Where the conditions are not of extreme severity, appropriate nutritional intervention can restore most if not all of the conditions of health and the continuation of the growth process.

There is one food that is clearly designed and balanced for human consumption. That is mother's milk. The question of mother's milk versus bottle feeding has taken on a number of political dimensions that have clouded the issue. The literature on the subject is massive and even the slightest perusal of it finds that the overwhelming evidence is that human mother's milk is superior nutritionally to the milk of other animals for feeding the human infant; it confers immunities upon the child; and it contains trace elements unobtainable in any of the powdered formula substitutes. Further, the circumstances under which poor people in third world countries use formula further contributes to its disadvantages as infant food. Poverty forces them to dilute the formula, and dirty water and dirty containers can lead to diarrhea and other infections. Oral rehydration therapy (ORT) for curing diarrhea using glucose, salt, and water shows that there are instances where the simpler technologies are sometimes the best. Clearly, infant feeding is one instance where nature knows best. The role of science and technology in infant feeding (at least for the first 6 months) is in overall nutritional enhancement for the mother so that she can manufacture it for the child. Further, it is a cliché but also a fact that infant nutrition begins in the womb. Therefore, a healthier, better-fed mother produces a healthier infant that is able to take advantage of the mother's milk.

The role of science and technology in world nutrition is multifaceted:

1. Scientific research will continue to expand our knowledge of nutrition and the basic nature of the human physiological life process, so that more effective strategies of nutrition intervention can be devised.

2. Science and technology can continue those processes that create land and enhance food output.

3. Science and technology can play a central role in

helping poor countries and poor farmers increase their food production or increase their income so that they may purchase adequate nutrition.

4. Science and technology, through fundamental genetic experimentation, can improve the nutritional quality of plants currently being cultivated.

5. Science can lead the way in exploring the rich storehouse of our biosphere for plant life with the potential to become cultigens and food sources, or for plant life that once was used as human foodstuffs.

Numbers (4) and (5) complement (2) in that they have the potential of allowing unused lands to be cultivated, or allowing more intensive use of existing land. Finding, creating, and utilizing new plants is the subject of Chapter 10.

CHAPTER 10
New Crops and
Novel Food Sources

PLANT BREEDING: PAST, PRESENT AND FUTURE
     Of the vast number of plant species that exist, humans have
used over 3,000 as food sources. As people have come into
contact with one another, the diffusion of crops from area to
area has caused some to supplant others as basic foodstuffs.
Greater contact, through improved means of transportation between
people and climatic zones, has meant an increased diversity of
foodstuffs to consume as well as more varied means of food
preparation. Though individual or group access to varied foods
may increase with contact and diffusion, overall the number of
plants grown or otherwise harvested has steadily diminished.
Currently about two dozen crop species provide the vast majority
of plant food upon which we depend.
     Norman Borlaug (1981) lists 23 plants that form the base of
world agriculture. They are

     1. Five cereals: wheat, rice, maize, sorghum, and barley.
     2. Three root crops: potato, sweet potato, and cassava.
     3. Two sugar crops: sugarcane and sugar beet.
     4. Six grain legumes (pulses): dry beans, dry peas,
chickpea, broad bean, ground nut, and soybean.
     5. Three oil seeds: cotton seed, sunflower, and rape seed.
The pulses ground nut and soybean are also oil seeds.
     6. Four tree crops: banana, coconut, orange, and apple.

     Over 80% of harvested food by weight is from plants, and
just about half of all food and calories are from the five cereal
grains (Borlaug 1981, 108-9).
     As domesticated plants were diffused to new regions,
varieties were bred and adapted to fit the environmental circumstances. Normal evolutionary processes led to a diversity of
other regional variations within the plant species. Much of this
regional diversity is now being lost, as local varieties give way
to the more highly productive, specially bred, high-yielding
varieties. It is true that most of the plants have been crossed
with local varieties to fit microclimatic and cultural needs.
Nevertheless, within some of the major crops, such as wheat and

rice, an increasing proportion of the world's output comes from plants that share a common ancestry. To some, this is a prescription for disaster.

Much of this concern is legitimate. Some, however, reflects opposition to the green revolution and is an outgrowth of anti-technology catastrophic thinking. Those involved in plant breeding programs are well aware of the need to maintain diversity. There is an awareness of the issue of genetic vulnerability in the scientific community. First, there is the much broader issue of diminution of species in both the wild plant and animal communities, as human settlements with seeming relentlessness convert habitat, such as rain forest, to cultivation. Apart from the aesthetic or ethical question (it is difficult to define or categorize it) of maintaining diversity on the planet earth, there are also direct practical reasons for doing so. Plant life continues to be a major source of new chemical compounds that have important human uses, particularly pharmaceutical. Many of these have not been replicated in the laboratory, or at least cannot be done so economically. It has been estimated that perhaps as many as 80,000 plants are potential human foodstuffs (Timberlake et al. 1982, 33). As argued in earlier chapters, the way to protect existing habitats such as rain forests is to improve the yield on cultivated lands.

More to the point of this chapter is the decline in genetic diversity of cultivated plants. In preserving "natural" habitat, included will be wild ancestors and relatives and closely related species of currently cultivated plants. This is an important part of the gene pool for current and future breeding programs (Coulter 1980, 39-40); and, of course, absolutely essential for breeding purposes are currently used cultivars (some of which may not be used very much longer) and varieties no longer planted. In recognition of the need to preserve this heritage, the International Board for Plant Genetic Resources was established in 1974 to coordinate the network of plant gene banks around the world. Various international and national agencies and research institutes are collecting and storing germ plasm for current and future use. Rice and wheat, as would be expected, lead in the number of accessions to worldwide collections. At the International Rice Research Institute (IRRI) at Los Baños, Phillipines, well over 60,000 varieties of rice are being preserved; furthermore, there is a worldwide search for approximately 30,000-40,000 additional varieties of rice under cultivation or in the wild (Swaminathan 1984, 85). Another source estimates that there are 120,000 rice strains in the world. In either case, IRRI plans to be storing samples of all of them by 1985 (Lewin 1982). At the reserve bank for wheat at Centro Internacional de Mejoramiento de Maiz y Trigo (CIMMYT), 90,000 entries are being held (Abelson 1982, 247). Duplicates of the major foodstuffs are kept in storage in other centers for backup protection. Duplicates for rice are kept in the U.S. National Seed Storage Laboratory in Fort Collins, Colorado (Swaminathan, 8). This facility is itself the center for regional, state, and local governmental and private storage facilities for plant germ plasm (Walsh 1981, 163).

Plant breeders in both the public and private sector are in the forefront of support for the network of storage facilities for seeds and germ plasm. However, many object to the argument that modern plant breeding has made world agriculture more vulnerable to disease and disaster. Not only are we gathering and storing these varieties of many different plants, we are also creating new varieties. Borlaug (1981, 163-64) argues that CIMMYT "makes a minimum of 10,000 different new crosses each year," and is active in wheat plant breeding in 100 locations in 60 different countries. The research agencies cooperating with IRRI request and receive about 5,000 to 10,000 packets of seeds each year and make about 5,000 new crosses (Brady 1980, 24, 125). A survey of plant breeders by Donald N. Duvick found that a "large majority of breeders said that the genetic base of their breeding programs has broadened since 1970" (Walsh 1981, 163). Not only is plant breeding creating new varieties, but it has also created a new plant, triticale, a cross between wheat and rye, which has been described as the "first nutritionally significant food crop to be introduced in the past 5,000 years" (Smith 1983b, 98).

As noted in Chapter 1, humans have had the remarkable achievement of spreading over the entire globe, without speciation, in the last 125,000 years. They have done this without the need for biological adaptation by using technology to adapt the environment and to create a niche within it for human habitation. Since domestication of plants and animals, humans have carried plant and animal life with them. The adaptation process for the domesticates has been one of both environmental and species modification. In some instances the idea of domestication (i.e., stimulus diffusion) has spread, with local plants being brought into production. Nevertheless, a small number of species, most of which originated in tropical or subtropical climates, are cultivated under a diverse set of macroenvironmental conditions.

Man depends on continuing supplies of fresh water for his own physiological functions. As a result, human populations have concentrated in areas of ample rainfall or river valleys that provide irrigation, such as the Nile. Consequently, domestication of crops has emphasized species adapted to soils found in such areas (Devine 1982, 143).

The very process of domesticating a plant is likely to involve both plant modification and environmental modification. Any form of field preparation, irrigation, etc., would be environmental modification and part of the activity that we describe as the creation of arable land. Fertilization of any kind would be environmental modification. Similarly, from hunting and gathering through domestication, to recent times, selectivity of plants becomes involved in plant evolution. Consciously or unconsciously, it is a form of selective breeding.

A form of stress, such as drought, that wipes out most of a crop
is likely to leave the most resistant strains for subsequent
planting. This form of plant evolution creates more resistant or
otherwise survivable strains. A farmer selecting out the "best"
seeds for next year's planting is engaging in long-term breeding.
Continued planting in an area, through plant or seed selection
and through random genetic change, gave rise to local varieties
that were suited to the environment. Even so, for the individual
farmer, environmental modification is part of the daily routine,
while breeding is a long-term process not always discernible
within a generation.

Throughout human history, but particularly in the 19th
century, when new areas were opened to cultivation, there was a
search for crops that grew in similar environments. Active
scientific research and selective breeding are largely a product
of the last hundred years, having gained particular impetus with
the rediscovery of Mendelian principles of genetics at the
beginning of this century. In many ways, the main thrust of
breeding was intensification of favorable traits and continuation
of trends that had existed under historic conditions of non-
scientific breeding.

Once agriculture is established in an area, environmental
modification is an ongoing phenomenon. Preparation and maintenance of the field are forms of environmental modification
carried out regularly. Virtually everything the farmer does is
environmental modification. Most of the organized activities and
research described in previous chapters, such as improving soils,
are forms of environmental modification. Opening new lands to
cultivation has generally been primarily a form of environmental
modification.

The green revolution was clearly a result of plant research
and breeding, but its triumph in agricultural production was
carried out in conjunction with the use of fertilizers, pesticides, and irrigation (i.e., environmental modification). The
potentials that are now being explored in basic plant research
shift the emphasis more toward modifying the plant, thereby
greatly reducing the need to modify the environment. For
example, one can lime acidic soils or one can breed crops
that are tolerant of them. Generally we have done a bit of
both. The environmental modification aspects of modern agriculture have involved large, intensive use of energy and other
resources. We have argued that we are not about to exhaust any
of these; nevertheless, some of them, such as energy, have become
more expensive and for others there are ancillary problems of
pollution. It is also argued that, though more potentially
arable land exists, the cost of making it arable is becoming
increasingly prohibitive. Thus, modern plant breeding can
produce the following types of benefits:

1. Continuation of the green revolution for improved
varieties that, coupled with environmental modification (i.e.,
fertilizers, irrigation, etc.) increase output.

2. Plant changes that substitute for environmental modification.

3. Plant changes that allow new lands to be brought under cultivation, which are not arable by current technology, or which are potentially arable but at a great cost.

Most plants currently under production are much more productive than their natural progenitors. "A modern cereal or fruit or vegetable may out-yield its wild relatives by 10, 20, even 100 times," writes Tudge (1983, 547), "and yet, if you look at the details of its growth, at the uptake of nutrients from the soil, or the capture of carbon from the air by photosynthesis, or the proportion of nutrients absorbed that is actually diverted into edible fruit (or leaf or seed), you find inherent and often spectacular inefficiency."

Improved plant efficiency is readily accessible through new techniques. Traditional breeding is limited to the package of characteristics possessed by the plants being crossed. Some of the possibilities of plant breeding today have the potential to directly alter the basic characteristics of the plant itself. Tissue culture is one of the promising methods. A small portion of a plant is placed in a growth hormone substance. A mass of cells, called a callus, is formed. The callus then can be broken into millions of individual cells. They can then be nurtured (in vitro) into identical, individual plants (clones) or subjected to further treatment. These cells can be subjected to the kinds of stress that inhibit plant growth, i.e., water or nutrient shortage, temperature extremes, or high salt or acidity. The surviving cells presumably would have greater tolerance. Adding stimulants to cause them to grow and divide allows the process to be repeated, intensifying the desired traits; lastly, they are separated to be grown into plants (Blair 1980, 12-15; Daly 1983, 18-20; DeYoung 1983, 53-59).

In Horizons, journal for the Agency for International Development, Rockwood (1983, 22) describes the most frequently used types of tissue culture:

"Callus culture: Masses of unorganized cell clusters, called callus, grow on the surface of a solid growth medium. Tissue sources commonly include embryonic, seedling and mature vegetative plant parts, or reproductive tissue such as ovules.

"Cell suspension culture: Plant cells and cell aggregates—usually from callus culture—are grown in a liquid growth medium that is agitated to provide aeration and break up cell clusters.

"Organ culture: Plant organs, such as roots, shoots, embryos, anthers, ovaries, and ovules are grown on either a solid growth medium or on a solid surface that is connected to a liquid growth medium by a wick.

"Meristem tip culture: Parts of meristem (shoot tips) are placed on a solid growth medium. Either single plants or large numbers of buds may be generated and transferred to a rooting medium.

"Protoplast culture: Protoplasts (wall-less cells in which the nucleus and cytoplasm are enclosed in the plasma membrane of the cell) are isolated from leaf mesophyll tissues, root tissue, or cell suspension culture. Protoplasts are cultured in a liquid medium--to allow protoplast manipulation and subsequent cell wall regeneration--and then transferred to a solid growth medium for callus growth and plant regeneration."

The use of tissue culture greatly compresses the time period for plant evolution and possible favorable mutation. It releases the grower from planting and raising crops, selecting varieties, and repeating the process. If a mutant plant with a bundle of desirable characteristics is produced by these methods, tissue culture can greatly accelerate the clonal reproduction and testing of the new variety. It saves greatly on the space needed to carry out research. According to Blair (1980, 15), "A single flask containing a pint of wheat cell suspension contains 40 million potential plants, or the equivalent of 3.2 hectares of field-grown wheat. A small room could easily hold 35,000 such vials. The researcher could therefore have at his fingertips the genetic equivalent of 112,000 hectares of wheat or 1.4 billion possible wheat plants." Tissue culture has the further advantage of allowing the development of disease-free clones that can be then sent out to different regions or countries without fear of spreading a plant disease (Rockwood 1983, 24). Tissue culture and other new biotechnologies serve as complements to more traditional breeding methods.

Another important technique of biotechnology is gene splicing or gene transplanting. In this technique a gene for a known desired trait would be taken from one plant, spliced into the DNA (deoxyribonucleic acid) of a vector and introduced into the cell of another plant. One vector that is being widely considered is the bacterium Agrobacterium tumefaciens, which invades plants and causes crown gall disease. An obvious candidate for this type of technique and one that is widely discussed is that of transferring to grain crops the capability of legumes to host nitrogen-fixing bacteria. Genetic engineering involves a variety of complexities, including not knowing how the implanted gene would act in the set of interactions of its new host. It cannot be stressed too strongly or too often that, though many of the promises of biotechnology are real, the difficulties of achieving them are also real. Thus, in a popular article the procedures of gene transplants may seem simple enough, but the actual carrying out of this effort is far more difficult. When one reads the literature on research to improve photosynthetic efficiency or rebuild and improve plants in other ways, projections of successful breakthroughs are frequently on the order of 15 to 50 years. Clearly, we all hope that these projections err substantially on the side of being too conservative. Even so, we still have the problems of the next few years to consider.

It is important to have coordinated research policies that strive for both long- and short-term gain. The Consultative Group on International Agricultural Research was set up in 1971 "to support a worldwide system of agricultural research centers and programs." It has helped to establish centers in different regions, and it has assisted already existing centers to incorporate some of the most advanced research techniques into their research program. Most of the gains in food output over the last decades have been in improved grains through use of fertilizers, pesticides, and irrigation. Consequently, the direct benefits (other than as consumers on the international market or recipients of food aid) of the green revolution have not penetrated many of the most needy areas of the tropics and subtropics. This is particularly true for large areas of Africa.

Support given by any group to tropical research institutes or to institutes specializing in crops other than grains (i.e., cassava or potatoes) is vital for agricultural progress in the third world. Currently, the United Nations Industrial Development Organization (UNIDO) is considering a biotechnology center oriented toward helping "third world nations develop expertise in biotechnology" (Dickson 1983, 1351-53; Seneviraine 1983, 50-51). Many of the people who strongly advocate such a center are molecular biologists who originated in third world countries but now work in industrial countries. One of the best ways to reduce the brain drain is to have quality research facilities in third world countries which give outstanding, well-educated minds the technical means to do first-class research. "By developed country standards," writes Rockwood, "the facilities and equipment needed for tissue culture are modest, but still may be beyond the means of many developing country researchers. And, in many countries, electric power supplies are still not dependable enough to assure temperature and light control."

Traditional breeding has brought agriculture a long way and helped to feed the world. Tissue culture, which is beginning to have its impact upon agriculture, offers enormous promise for the future. Gene implantation to date is more promising than it is real. There is a continuing need for crossbreeding. This will require that we maintain and increase the large banks of genetic stock. Such banks are useful not only for crossbreeding; eventually they will be the source for gene implants. The discovery of colchicine in the 1930s has greatly assisted plant breeding. Many crosses are sterile. Colchicine causes the haploid to split into a diploid or a "dihaploid" that "is perfectly homozygous." This also produces uniformity for breeding. The use of colchicine was essential for the creation of triticale (Smith 1983b, 95). It is widely used today for various breeding programs.

Considerable work is being done in crossing barley with hardier
wild varieties (Tudge 1983, 548).

The use of wild varieties for breeding, the potential
crosses between different varieties, and the use of genes from
other plants in genetic engineering all make it worthwhile to
survey some of the plant potentials. Some of these plants are
also potential domesticates in their own right.

WILD PLANTS: A FEW EXAMPLES
In 1976, the National Academy of Sciences (U.S.) published
Underexploited Tropical Plants with Promising Economic Value.
Thirty-five plants that were either not exploited or used only in
limited regions were discussed. There has been an enormous
demand for the book, and several reprintings were exhausted.
Separate monographs and studies have been carried out on the
potential of several of these plants. There has been a veritable
flood of popular articles and newspaper accounts on the exceptional virtue of some of the plants. With the exception of
leucana, most of the miracle foods discussed in this section were
studied in Underexploited Tropical Plants. The obvious exag-
geration of some of the popular accounts may cause some observers
not to realize that there are substantial potential benefits in
these plants if we understand their strengths and weaknesses and
engage in programs for plant selection and improvement.

The following is a brief survey of a few exotic plants.
Some are already in use on a regional basis. Some are currently
being experimented with in different parts of the world and
at research stations. Included are references to some wild
varieties of domesticated plants. These can serve as a source of
genetic characteristics for crossbreeding purposes. In the long
run, with gene splicing, all of these plants and organisms will
be part of the biological heritage for sophisticated plant
creation. All merit continued study and experimentation. None
is likely to provide a miracle cure for our human food and
malnutrition needs. If any turns out to become one of the 20 or
so leading foodstuffs, that would be a bonus; certainly it is
possible for any or all of them to make important contributions
to human needs.

Amaranth and Quinua
Amaranth was a major grain of the Aztecs. Some argue that
it was deliberately suppressed by the Spanish conquerors because
of its association with indigenous religions. There are 3
species of amaranth that are viewed as a potential protein
source. Breeding work with colchicine has already been done to
create more heat- and drought-resistant varieties. Amaranth is
high in lysine, which makes it a complement to other grains, such
as wheat, for which lysine is a limiting amino acid. There is
currently some cultivation of amaranth in south and southeast
Asia, as well as in the southwestern United States. Amaranth is
most likely to be significant for its nutritional qualities,
rather than for its yield or for its ability to grow in marginal

environments (National Research Council 1975b, pp. 14-16, 1984; Weber 1978, 3-5; Tonge 1979; Vietmeyer 1981, 709-12; Brody 1984).

Quinua, another "lost grain" of an ancient civilization (the Incas), is also seen as having potential either for cultivation or as a breeding stock. Quinua is a hardy plant that grows in higher elevations in Bolivia, Peru, and Ecuador.

Corn (Maize)

Corn is one of the 3 major grain crops in the world, along with wheat and rice. In less than half a century, corn yields per acre have tripled. Cross breeding with varieties from around the world is considered an important factor in hybrid vigor and increased yields. There is a recognized need to sustain inputs of new sources of germ plasm to maintain this vigor. Crossing with teosinte (a wild plant and possible progenitor of corn) is considered to have potential improvement. It also holds the possibility of making domesticated corn a perennial. This could save on fuel and other costs in field preparation and planting in a regimen of minimum tillage (Cox 1979, 40; Villareal 1980, 1-10; Myers 1981, 70; Brynka 1983; Scott 1983, 49-50).

In earlier research on corn, a mutant gene (opaque 2) was found, which caused the corn to have twice as much lysine and tryptophan, 2 limiting amino acids. Though the variety is undesirable in other respects, this mutant gene is likely to be used in future breeding programs.

Guar and Guayule

Guar and Guayule, though not a source of food, as are most of the plants discussed here, are potentially important cash crops for farmers in arid and tropical regions. Guar produces gum, which is used in a variety of processed foods and has a variety of other industrial uses. Guayule is a potential source of rubber that, when processed, is identical to that from the Hevea tree. Because guar and guayule grow on poor sandy soils, are drought resistant, and can stand high temperatures, they are potential crops in arid lands where virtually no other cash or food crop can be grown (National Research Council 1975b, pp. 145-49, 1977a, 1979, pp. 4, 279; Wilford 1980).

Jojoba

Jojoba is one of the plants most frequently described as having almost miraculous potential for human well-being. The other two that have caught the popular media's attention are leucaena and the winged bean. For jojoba, some of the promise may become reality. It is hard to visit any of the drier regions of the tropics or subtropics without finding experimental efforts at growing jojoba. Some of these are fairly large-scale projects. Clearly the agricultural scientific community, aid donors, various international and domestic agencies, and the private sector think that jojoba has important economic possibilities (National Research Council 1975a; Huang and Mayfield 1980, 95-96; Robinson 1981; Vietmeyer 1981, 704-7).

Jojoba is a desert shrub found growing wild in the Sonoran Desert of northern Mexico and the southwestern United States. It produces a very high-quality oil with many uses and a high market value. For instance, jojoba can be transformed into motor oils, rich creams that smooth and stabilize cosmetics, and hard, sparkling waxes. It provides an excellent substitute for oil from the sperm whale. The average jojoba bush produces about 5 pounds of seed, and these seeds are approximately 50% oil.

Jojoba is usually bushy and can grow as high as 10 feet. It has a life span that is estimated at between 100 and 200 years. It tolerates extreme desert temperatures and daily highs of 35-45 degrees centigrade. It is drought resistant and thrives under soil and moisture conditions not suitable for most agricultural crops. Productivity is increased when annual rainfall is about 15-18 inches. However, rainfall as low as 4 inches a year may produce seed. The shrub has been known to survive as long as a year with no rainfall at all.

The plant has some potential for browse for animals. The oil seedcake is high in protein and has potential for animal feed if toxic substances can be removed. It would appear that there is solid economic potential for jojoba for agriculture in dry, hot areas throughout the third world.

Leucaena

Leucaena is one of a number of leguminous, fast-growing trees that serve a variety of useful purposes in third world countries. These trees can play a vital role in reforestation and land conservation throughout the world. Being leguminous, they can help in regenerating agricultural lands with nitrogen. Being fast-growing, they can provide for harvesting of fuel and forage. However, with the continuation of this latter use, though the trees may be adding nitrogen to the soil, they also will be depriving the soil of other important nutrients. Continued utilization of leguminous trees requires maintenance of the soils, just as in other forms of agriculture (National Research Council 1977b; Aziz 1979, 16-17; Benge 1983, 31-33).

Leucaena can be found as either a shrub or a tree. In shrub form, leucaena can be grown as a high-protein forage. It also can be harvested and fed to animals. It has, in addition, innumerable other potential uses as human food, fertilizer, and fuel, so that it has been referred to as a supermarket tree. The use of such hyperbole to describe leucaena points to some of the dangers involved in the advocacy of "miracle" plants. Michael Benge (1983) writes, "While many fast-growing trees generally have done well, poor performance in many areas has severely damaged their credibility. The 'miracle' is questioned by many who unquestioningly listened to the early advocates of these trees who may have overstated the case in their enthusiasm for the potential they seem to have."

The kinds of difficulties that have caused leucaena (and other fast-growing trees) to lose their credibility can be a key to understanding the problems and potential solutions of other nontraditional plants. First, says Benge, basic research and information about these trees is "woefully inadequate . . . despite an investment of hundreds of millions of dollars." What is known is "not widely disseminated, evaluated, and adopted for local use." The causes of the failures are many and varied. For leucaena, there has been "poor performance and great variance in growth due to varietal impurity, adulteration of the seeds, and poor selection practices." There are also difficulties for nonruminants in digesting the leucaena leaves.

Many of these "miracle" plants are promoted by some who are suspicious of modern science and technology and see it as the cause of our problems. Ironically, if plants such as leucaena are going to come anywhere near fulfilling expectations of them, it will be because of the breeding techniques described earlier in this chapter. Leucaena is already fairly widely used in Asia, Central America, and Hawaii. Evidence is that in some instances, its use may extend back 2000 years. Rather than viewing leucaena in terms of varieties of shrubs and trees, it would mainly be seen as a storehouse of very useful germ plasm. Tissue culture and other plant genetic techniques will be necessary to select, develop, and propagate improved superior varieties. Breeding techniques will be useful in adapting leucaena to local conditions and needs. Scientific understanding of such factors as nitrogen fixation, rhizobium, trace elements in tropical soils, and the role of other organisms in the soil (such as mycorrhizal fungi) will be necessary for the proper cultivation of leucaena. Benge (1983, 31), one of the strongest supporters of leucaena and the one most often cited in accounts, makes the following observation:

> Until trees such as leucaena and the other fast growers are treated as a crop with emphasis on genetic improvements, breeding, nutrient needs, pest control, and proper management techniques, the opportunity that they could be miracle trees will be lost. They will be merely marginal trees on marginal land, contributing very little to efforts to provide increased food, fuel, fodder, and wood to the rural poor.

Most scientific works in the "miracle plants," such as those by Benge, Noel Vietmeyer, and other authors writing for the Agency for International Development or for the National Academy of Sciences (U.S.), recognize that there are serious limitations and problems in the use of these plants. All of their work cited here has sections on the limitations or constraints. It is only by understanding these constraints that the research agenda can be established to overcome them and provide the basis for reaping the plants' benefits. This is true for leucaena and for other plants surveyed in this chapter.

Morama

The morama bean is the only plant discussed in this section that has not been or is not currently being cultivated by some group of people. However, as a wild plant it is an important component in the diet of the Khoisan people of the Kalahari Desert (National Research Council 1979, 68–74; Bousquet 1983; Nicol 1983).

The nutritional value of the morama bean is described by the National Research Council, Academy of Sciences (1979, 69–70), as follows:

> The morama bean analyses so far reported record protein contents of 30, 34, and 39 percent, respectively, a range roughly comparable to that of soybean, 37–39 percent. Oil content is reported as 36–43 percent of the dry seed by weight. Thus, the marama seed has a protein content that rivals that of soybean, approaches that of the peanut. The seeds have less than half the fiber of peanuts and are a source of nutritionally important minerals. Like most legume proteins, marama bean protein is rich in lysine (5 percent) and deficient in methionine (0.7 percent).

Our limited knowledge of the morama bean is the greatest limitation to its expanded use. We know it has some potential, but we really do not know how useful it could be. Given the poor soils, the paucity of water, and the temperature stress of its natural environment, the morama bean is likely to be cultivable in what are now marginal and uncultivable lands. Being leguminous, it should have less need for fertilization. All of this is contingent on continued research and experimentation. Some work has already been successfully carried out on marginal land in west Texas (Bousquet 1983).

Neem

Neem trees have a long history of diverse uses in South Asia that continues to the present. In India, neem trees are grown for fertilizer, timber, and as a source for skin ointment and toothpaste (Moffat 1979). Neem trees also make good shade trees. Strips of wood from the neem tree are also used for brushing one's teeth. One author (Saxena 1983, 147) notes that "neem is relatively non-toxic to man and other vertebrates. Neem oil is used to make soaps and detergents, and neem derivatives are used in toothpaste and herbal medicine. Neem cake has been used as a feed supplement for cattle, livestock, and poultry."

Neem trees have attracted international scientific interest because of the effective natural pesticide that can be extracted from its seeds. Plants have been in a continuous battle with plant-eating insects as long as the 2 have existed together. One would expect that some plants would evolve defense mechanisms against insects of various kinds. The seeds and leaves of the neem tree yield azadirachten, "a compound that appears to be a promising new insect repellent" (Brady 1982, 851). The people in India have long been known to mix in a few neem seeds with their stored grains for insect protection. In the search for natural pesticides in plants, neem is certain to play an increasingly important role. The International Center of Insect Physiology and Ecology (ICIPE) in Nairobi, Kenya, is leading in the study of natural pesticides. There has also been an international conference to explore the potential of neem.

Potatoes

The potato, like corn, is a major foodstuff, for which it is believed that cross-breeding with wild varieties can bring considerable improvement. It is also a plant where there is concern about the diminishing genetic variability. In Europe, 1 species of potato encompasses all the varieties cultivated (Smith 1983a, 558). In the United States, nearly 40% of the potatoes grown are of one variety, Russet Burbank (Shepard et al. 17-24). In contrast, in South America (where the potato was domesticated), there are 8 species cultivated, and within these are large numbers of landraces. One anthropologist actually counted 46 varieties in one field in Peru; varieties steadily are yielding to a much smaller number of higher-yielding types, so that the rarer varieties are now found only in out-of-the-way places, particularly in higher elevations on mountains (Smith 1983a, 557-58). As with other major crops, germ plasm banks have been established to preserve the basics of genetic diversity. Worldwide, there are about 44,000 potato accessions (including duplicates). Centro Internacional de la Papa (CIP) in Lima, Peru, has 13,000 accessions. These include varieties of all 8 domesticated species and a large proportion of the 154 wild species (pp. 559-60). Plant characteristics are recorded on a computer (up to 54 per accession). This system is used for identifying and retrieving seeds for shipment to other breeding centers.

Crossbreeding with rare cultivated varieties and with wild species offers man opportunities for improvements in growing potatoes. Particularly important is the ability to withstand frost and disease. Preharvest loss of output due to disease or frost is high for potatoes relative to other cultivated plants. About 22% of the crop is lost to disease alone (Shepard et al. 1980, 18).

The potato is a plant for which the techniques of tissue culture are currently workable. That does not mean that breeding programs of this type will necessarily be successful, but the potential is there. Plants have been successfully regenerated from leaf-cell protoplasts, and researchers believe that this research can generate improved lines.

Spirulina and Other Unconventional Foods

N. W. Pirie (1976, 64) estimates that 99% of our plant food comes from species domesticated 5,000-10,000 years ago. He argues that we have to stop assuming that ancient cultivators had unsurpassable culinary wisdom and start looking for other sources of food. In short, we have to find uses for what already grows well. Not only does Pirie explore the use of novel plants, but also other types of organisms, such as unicellular green algae. Spirulina is a blue-green algae that was consumed by the Aztecs and is considered by many as an important new foodstuff. It grows naturally in saline and alkaline lakes where its entangled manner of growth makes collection easy. It is a traditional human foodstuff in places as diverse as the Lake Chad region of Africa and Central Mexico, and it is used as animal feed in Japan. It is high in protein and "contains methionine, tryptophan, and other essential acids, at concentrations similar to, if not higher than, those present in casein--the chief and unique protein of milk" (Ciferri 1981, 811). Its potential dry weight yield in protein is extraordinary, easily exceeding its nearest competitors. After all, much of our breeding of dwarf and micro-dwarf plants is to decrease the lower-valued vegetable structure and increase the relative size of parts that are basic foodstuffs. In the microbial organisms we have an entire organism that is the foodstuff (Furst 1978, 60-65).

Spirulina, and most of the algae, are efficient photosynthesizers. Many of the plants that we have been touting are useful because they grow on highly acidic soils. Spirulina grows in highly alkaline water with a pH as high as 9.0. It can also be grown on sewage and other kinds of waste. A variety of other microorganisms can also grow in these environments and provide additional service, such as consuming waste products. Many readers who may not be interested in eating fungal foods should realize that included in this category are mushrooms and truffles. Microorganisms already perform a variety of human services, such as helping preserve food or making it more palatable through fermentation. Leaf protein and other foodstuffs manufactured from existing wild plants also have considerable potential as unconventional foodstuffs.

Spirulina and similar organisms as valued sources of protein and food have been known for so long that it is obvious that there are cultural and technical barriers to their expanded utilization. It is comparatively easy to write articles or sections of books on the advantages of a foodstuff. This is a right and necessary step in identifying, understanding, and disseminating knowledge about agricultural potential. The next step, however--finding solutions in the field--is still the most difficult.

Triticale

Triticale may be a peek into our agricultural future. It is the first and essentially only man-made crop. It is the first new cultigen in several thousand years. Triticale has been around in an experimental stage for several decades. It made the transition as a commercial plant at the end of the 1960s. Its growth in output since then has been spectacular in percentage terms, but, starting from a low first year, rapid growth still has not led to a large absolute output.

Triticale, a cross between wheat and rye, has a more complete amino acid chain than either of its parents. It is more nutritious than any of the grains. However, its yields are generally lower and there are still problems in seed reproduction that would be expected from a cross between the two species. Triticale seems more tolerant of aluminum toxicity in soils and has outperformed some of the wheats in Brazil that were bred to grow in some of the highly acidic tropical soils in the Campo Cerrado. The steady growth in output indicates that triticale's potential is realizable. Though it might not surpass any of the 3 major grains in output in the next few decades (in fact, it is highly unlikely to do so), it still could be one of the 5 or 6 major grains, particularly in the tropics.

Ironically, in the United States this man-made grain is found mainly in what are called "natural" food stores. When mixed with wheat, which is higher in gluten, it makes an excellent bread, as those of us who have become devotees of it will readily testify. Of course, this is no different than some of the other grains, for example, oats, which have to be mixed with wheat to make Western-style bread. Alone, triticale can be prepared in many of the same ways as other grains. Experimental work is currently under way in crossing other grains, particularly wheat and barley. A variety of characteristics are being sought, including greater resistance to disease and other forms of stress.

Winged Bean

Of all the "miracle crops," the winged bean is one of the easiest to become genuinely enthusiastic about. It is not uncommon for public lecturers on economic development to receive questions from the audience about the winged bean. At times they are almost conspiratorial in tone, inquiring as to the reasons it is not being more widely promoted. Though it may not have all the possibilities of saving humanity from hunger, as implied by the questioners, it might supplant soybeans as a source of oil and vegetable protein. If we look at the very rapid rate of increase in soybean production over the second half of this century, we get some idea of the real possibility that the winged bean could become a major food crop.

The winged bean is a tropical legume with a multitude of exceptionally large nitrogen-fixing nodules. It produces seeds, pods, and leaves (all edible by humans and livestock) with unusually high protein levels, tuberous roots with exceptional amounts of protein, and an edible seed oil. . . . It is a fast-growing perennial that is particularly valuable because it grows in the wet tropics where protein deficiency in human diets is not only great but difficult to remedy (National Research Council, 1975a, 56).

The vast array of uses of the winged bean has also led to its being called "a supermarket on a stalk." It is still primarily a garden crop in Papua New Guinea and Southeast Asia. Far too little is known about it: "What is known today about the winged bean is roughly tantamount to what was known about the soybean 60 years ago, shortly after its commercial introduction

to the United States. Many advances in the genetic improvement
of the soybean could be applied to the winged bean" (National
Research Council 1975b).

     Preservation of the planet's genetic diversity is essential
for sustained agricultural development (Daly 1984; Hahn 1984).
Much of the genetic material that has been vital to the growth of
agriculture in industrial countries has come from plants in the
third world. To some, the drain of genetic resources and the
increasing concentration of control over seed distribution by a
few multinations is a threat to continued third world survival.
It is argued that these countries are losing sovereignty over
precious resources (the "gene drain"), and that the well-being
of the entire world population is threatened by a narrowing
genetic base in agriculture (Mooney 1983, 1-2; Walsh 1984,
147-48). Further claims are made that one-half to one-third of
the gene samples in storage are lost to spoilage (MacKenzie 1983,
870-71). The genetic resources of this planet do have to be
used cooperatively if maximum benefit is to be achieved from
scientific research in agriculture. Agricultural research
cooperation and germ plasm storage is only part of the story.
     Earlier chapters have tried to counter the claims of the
doomsayers with evidence that genuine progress has been made.
The task of this chapter is almost, but not quite, the opposite.
Though not denying the enthusiastic prognosis made for the
benefits of new techniques of plant breeding or new plants, we
are trying to temper this optimism with a more realistic assessment of both the tremendous possibilities and the considerable
difficulties involved in achieving them. Nevertheless, if 50
years ago someone had predicted the gains that would be achieved
in the production of corn, they would have been thought to
be wildly utopian. If 25 years ago the gains of the green
revolution had been forecast, a similar critical assessment would
have been made. For that matter, in the entire range of rapid
change over the last 30 years or more, from decline in infant
mortality and increased life expectancy to gains in per capita
income, particularly in places such as Korea or Taiwan, anyone
foolish enough to have seen this future would have been deemed
unhistorical, unscientific, and not in complete command of their
senses.
     In many ways, the techniques that we are discussing
are far more spectacular in potential than those that have
brought us this totally unpredicted pace of change since 1950.
Realistically, the scientists doing the research see difficulties
as well as promises, and they essentially seem to be suggesting
that the near future rates of change are likely to resemble those
of the recent past. For policy purposes, research in biological
breeding techniques, seeking out new plants, and preserving germ
plasm should have high priority; they are probably currently
underfunded. Past gains and conservatively understood future
gains in agriculture more than warrant this judgment. It is
almost a cliche in the literature that agricultural research has

had the highest and fastest payback of any form of investment. This is merely measured in pecuniary terms and does not take into account the well-being of better-fed people. Thus, based upon very restrained forecasts, it is wise to continue to expand the activities described in this chapter. The incurable optimists, however, can only hope that future rates of change might be even greater than past ones. Overly enthusiastic selling of a technique or the potential of a plant is dangerous for policy decisions. If we have the right policies, however, maybe, just maybe, the enthusiasts might have just a kernel of truth.

PART IV

Technology in Theory and Practice

CHAPTER 11

Technological and Valuational Processes

The core of the definition of technology used throughout this book is that of a problem-solving process. We have argued that the evolutionary process of technological change is not inconsistent with the evolutionary processes of life. In fact, there are remarkable, but possibly superficial, similarities between them. Insofar as technology is a problem-solving process, it is a life-sustaining and -enhancing process. Problem solving involves people-choosing, an endeavor that is valuational. There is clearly a definable process by which reasonable, knowledgeable people select the correct tool to solve a mechanical problem. Questions of taste hardly matter in the determination of the size of wrench that one uses. This method of thought has penetrated other realms of human endeavors. As Veblen ([1906] 1961, 17) said, "Hence men have learned to think in the terms in which the technological processes act."

THE UNIVERSALITY AND PARTICULARITY OF TECHNOLOGICAL VALUATIONS

Technological values are what we may choose to call the valuational problem-solving activity implicit in technology, but these values are involved in other areas of human endeavors. They are, however, most easily observed and defined in the study of the use of technology and in the activity of scientific inquiry. Being human values, they derive their validity only to the extent that they promote the life process. As argued in a previous chapter, technology and technological values do not destroy cultural values but instead provide the means to find new ways of expressing them. Just as technology is both universal and particular, so are technological values. Problem specification in the use of technology for development involves both the particular problems of the physical environment and the cultural perception and definition of them.

The universal particular distinction has another dimension in terms of technological values. A group or individual can have a particular set of values, some of which are, at the given time, outside the possibilities of a technological valuation or scientific valuational test. A preference for fatty foods is not subject to any external test as long as there are no known consequences of that preference. When knowledge of nutrition allows a causal judgment to be made--that consumption of such foods beyond some minimum is life-threatening--then technological valuations become relevant. To argue against the purported causal relationship other than on scientific and technological grounds is a display of ignorance. The imperatives of judgment based upon scientific and technological consideration can become "morally and legally obligatory." Failure of a person operating in a professional capacity to use scientific standards in carrying out a job, be it medicine, law, finance, or bridge building, can result in civil and/or criminal liability (Haskell 1984, x).

To state a preference for a shorter life with fatty foods over a longer one without them is to express a taste. Taste, then, can be defined as a preference either that has no consequence other than to those who express it, or that has no known consequence at all beyond the immediate experience. Thus defined, we are trying to avoid several potential pitfalls. One is an extreme cultural relativism that prevents any cross-cultural valuations. Few would accept racism or any other form of bigotry as being merely the expression of a culture. These "values" have consequences that civilized people reject. Frequently they have been defended in terms of some alleged scientific theories proving the superiority of one group and the inferiority of another. Insofar as this is the case (and it is frequently the case that myths simulate the methods of science and technology), then they are subject to falsification by the processes of scientific and technological inquiry. One of the distinguishing characteristics of a scientific or technological theory is that it can be subjected to tests for falsification.

Tastes that have consequences only for those who express them do not necessarily carry normative prescriptions for change when historic, implicit, causal connections are falsified. However, when tastes are demonstrated to be life-threatening (e.g., eating fatty foods), then in the modern world at least there does appear to be a response in the form of a movement away from them. The influence of knowledge processes in behavioral processes sometimes seems painfully slow, as is evidenced by the number of people still smoking after overwhelming evidence of a causal connection to lung cancer and heart disease. In other instances, in countries like the United States, the response is at times premature, as the announcement of a possible adverse causality brings instant panic. Our distinction also seeks to avoid the pitfall of treating these tastes at the extreme either of a nontestable preference or of viewing the falsification process as an absolute moral imperative. There is a difference as to the falsification of a theory of social prejudice and the falsification of less consequential cultural norms.

Tastes that have no known consequence are outside the realm of technological valuations but nevertheless are not inconsistent with them. Those are people's preferences for certain tastes in food (such as hot, mild, delicate, and so on) or for colors or any other styles and attributes that delineate the cultural

distinctions of a people. Technology gives people new means to
express these cultural distinctions. Defining these values as
distinctive is another way of saying they are particular. Yet it
is the universal values that provide the means to protect and
expand these particular values. The workings of these principles
can be illustrated with the idea of nutrition for longer life.

Nutrition for longer life is essentially a scientifically
definable concept. We recognize that there are nuances in the
way culture defines good health or the cultural context in which
food consumption operates. Nevertheless, one can discuss certain
basic nutritional "needs," and in so doing one posits a universal
need, if not a value. Taste is a particular value. As new
foods become available, via technology or other means, cultural
taste will be a determining factor in their selection and
preparation. Given the widening range of choice that technology
facilitated in foodstuffs (and in most all other areas), should a
cultural preference for a given taste be shown to be harmful,
there are now far more options for realizing a semblance of
these taste preferences in nonharmful ways. People can alter
their diets or change the mix in the food consumption without
necessarily abandoning the distinctiveness of their cuisine. The
technology that is expanding food production is contributing to a
lengthening of life; scientific and technological valuation
processes also contribute to the extension of life in the same
way, both in the knowledge of nutritional causal relations and
the greater range of choices.

In summary, then, technological values are human values
and because they are based on causal relationships to the basic
human condition, they are universal. Technological values,
like technology itself, are also particular, because in their
problem-solving character certain problems are specified as
particular to a given culture and location. The causal relationships involved in the universality (and to some extent the
particularity) of technological valuations allow for these
valuations to apply to tastes that have empirically verifiable or
falsifiable consequences. Particularity of tastes that give
distinctiveness to peoples and cultures can be enhanced by the
expanding means offered by the dynamic universal technological
process of change.

The technological valuations in problem solving, like technology, language, and other processes now being described, are
open-ended. There is no teleological movement to a foreordained
perfect end. It is a process of bettering and improving, not of
perfecting in the strict sense of that term. Because there is no
ultimate end, only ends in view, it does not mean that the
process does not have direction. This is essentially the case
with all processes described here.

The important difference between technological evolution
and other evolutionary processes is the purposefulness implicit
in technological valuations. Early in the process people were
acting intelligently (i.e., purposeful, goal-oriented behavior)
for short-term objectives that had unintended long-term, technological consequences. As the process continues and becomes

self-reflective (just as language is), then the time horizon for purposeful, goal-oriented change lengthens. Today we are doing our best, however feeble our best may be, to look at scientific and technological possibilities far into the future. We are looking and seeking to control. Command over the direction of technological change needs a common set of values so that technology remains subservient to the needs of all, not just some, people.

Problem solving, then, does not mean the end of problems. New technologies solve old problems and create new ones. The total processes of life create new problems. Technological progress can be understood to have occurred if the problems created by technology are less harmful to the human enterprise than those solved. We should not be surprised that the processes of technological and social evolution are imperfect. This is just another way in which they resemble biological evolution (Cherfas 1984, 28-30). For the solution of these human problems, we continue with the technological process that solved their predecessors. This leads to the statement that is considered thoroughly outrageous by the critics of technology and that puts the utterer far beyond the pale of any possible redemption. In fact, it is frequently stated as a supposed caricature of the protechnology position. Namely, the solution to the problems created by technology is more technology. A few years ago, in response to an article of mine, I was accused of suggesting that technology was unique in providing the solution to the problems it created. My immediate emotional response was, Heavens, yes! My actual written reply was more prolix but possibly less to the point (Hayden 1980, 211-19; DeGregori 1980, 219-25).

In terms of the technological values we have been endorsing, the test of whether technological change has indeed been technological progress is whether our problems are worse today than in previous, supposedly idyllic, times. Obviously the total test of this proposition in written form would occupy many volumes. In part, other chapters have already addressed it in arguing for support of technological change. The final chapter makes a brief attempt to test this proposition against what are considered some of the worst features of modern life, such as pollution.

Technology and Institutional Adjustment

The translation of technological valuations into institutional practice is not always easy. Customary beliefs and practices do not always give way easily to the imperatives and demands of new circumstances. In recent years the maladjustment between technology and institutions has been blamed on technology. The implication of much of the criticism is that the technology will not work under any possible set of institutional relationships.

Technology has increasingly given humanity power over the environment. This has been translated into personal command. In the affluent societies, the individual is far more the lord of his (or her) castle than was any potentate of previous societies. Affluent societies have created a shelter against all but the rare and extreme ravages of the environment. The very technological processes that created and decentralized this power to the individual, however, at the same time centralized and globalized the consequences of the use of this technology. Until the development of flues and their stoves in Europe a half millenium ago, the vast majority of people of the colder climates heated their indoors with open fires. This filled the abode with smoke and particulate matter, which is carcinogenic. Those who have worked in development in areas where women work over open fires know the high incidence of eye and respiratory disease that follows. Furthermore, open fires are a threat to young children, who could get burned, and to the dwelling, that could go up in flames. Not only that, open fires were not an effective means of warmth, though they were the only means. The radiant heat of the fires toasts the side of the person facing it, while the drafts and cold, ambient air chill the side away from it. The "oil crisis" and the switch to wood-burning stoves in the United States has led to increased loss of life and property because of fires. The romanticism of the crackling fire is an affectation most easily enjoyed by the affluent, who have other heat sources. In fact, an open fireplace in a centrally heated home will draw more warm air out of the room and up the chimney than it will add heat to the room. The specially modified fireplaces with glass doors and special venting can add radiant heat at the cost of most of the crackle and at still some risk of fire.

The attempt to use biomass (such as wood) as a means of home heating has also been shown to be environmentally destructive in the creation of the fuel itself. It would require enormous amounts of land dedicated to biomass creation (Calef 1976). To use other forms of renewable solar energies also involves pollution and other environmental hazards. Active solar systems use toxic substances in their creation, operation, storage of energy, and maintenance of the system. There are the problems of the structural weight of the system, leaks or corrosion of toxic fluids, and the toxic wastes that have to be disposed of in cleaning the system (Bossong 1979, 4-7).

In the use of windmills, there are problems of noise, dangers from loose, flying parts during high winds, and danger to low-flying birds (Bossong 1979, 6). In addition, for most active solar systems there are the requirements for stringent environmental controls, from preventing the growing of trees to limiting the height of structures, in order to protect access to the energy source, be it wind or sun. As one solar advocate stated it, "The power of the sun is awesome. But it can be blocked by something as seemingly fragile as a leaf. If we are to harness solar energy, barriers between the sun and solar collectors must be prevented (Buckley 1979).

For passive solar systems (that is, tightly sealed, well-insulated structures), the problems of indoor pollution are once again becoming significant.

The mention of indoor air pollution often provokes reactions of disbelief or humor or both. Although many

questions remain to be answered, scientists are becoming
increasingly convinced that indoor air pollution is a
serious health problem. The indoor air pollutants now being
studied are radon and its decay products, chemical products
of combustion such as nitrogen oxides and carbon monoxide,
formaldehyde, asbestos, residues from consumer products,
allergens and micro-organisms, and tobacco smoke. These
pollutants can cause respiratory diseases, cancer, and even
death.

Indoor air pollution may, in fact, pose an even greater
threat to health than outdoor air pollution. Contrary to
common assumption, air pollutant concentrations behind the
closed doors of homes and other buildings are often higher
than corresponding concentrations outdoors. This is
particularly true of pollutants such as formaldehyde that
are released primarily indoors, but it can also be true of
pollutants such as nitrogen oxides that are released both
indoors and outdoors.

Moreover, most people spend most of their time indoors.
Thus, even if pollutant concentrations are lower indoors
than outdoors, indoor exposures are more prolonged and
frequent. Consequently, health effects may be more severe.
Although indoor air pollution affects all groups of people,
it particularly threatens the young, the old, and the ill;
these groups are both more susceptible to the effects of
pollution and more likely to be indoors (Kirsch 1982,
339-40).

The literature on indoor pollution is massive and growing.
Kirsch cites a large number as sources for the above paragraphs
alone. In addition, there are a number of popular scientific
articles on the subject. This does not mean that greater
insulation may not be cost-effective as a form of energy
conservation, nor that some of the hazards to air quality cannot
be reduced with the use of complementary technologies. It does
mean that life--from shelter to nutrition to whatever--entails
risks and problems, and there are no magic risk-free, problem-
free solutions. There are ways of doing things that are better,
and that is what technology and technological valuations are
about.

For those who can afford it, technology has solved many of
the micro-environmental problems, such as shelter, but in doing
so it has added to larger-scale problems. The affluent may
complain about polluted water, but they are not the ones drinking
the water that may account for as much as 80% of the communicable
diseases in many parts of the world. In early Egypt, at least
some people took in highly polluted air, as evidenced by the
carbon in the lungs of a mummy that was studied (Eisenbud 1978,
17). In the medieval European village, the actual indoor
breathing atmosphere was more polluted than modern urban society
would tolerate for its environment. One can even imagine in

colder weather, with plenty of fires roaring, a pall of heavy
pollution hanging over the village itself. In the scattered,
less densely populated societies, relative to today, the
inadequate technology of fuel consumption almost certainly meant
that one village's air pollution did not adversely affect
another's. However, one village's water pollution could in fact
affect those downstream if the stream was used as a human waste
dump.

Harvey Brooks argues that modern technologies are not in
themselves more polluting. Per unit of output, modern technologies tend to be less polluting than earlier technologies.
Electric generation for light and for other energy uses, such as
heating or power, is less polluting than candles or wood stoves.
The automobile is less polluting and uses fewer resources than
does the horse and buggy. The pollution of modern times is
derived from the scale of output and therefore consumption of
these services (Brooks 1980, 71).

Modern technology, then, is power: power on a scale
unprecedented and unimagined in human history. It is the power
both to do good and to do evil. This includes an intentional
evil. Pollution, species extinction, and the like, are not new;
it is the magnitude of their reach that is. The power that
technology confers upon the individual has global consequences.
The exhilarating experience of Western individualism has not
always been tempered by the commensurate recognition of human
rights. The individual who polluted his residence because of the
need for heat did little damage to his neighbors (except if his
dwelling caught fire and the fire spread) and certainly none to
distant villages. Power was limited, and there was no need for
institutional control.

We have not fully made the institutional adjustments needed
to manage our technologies. "Not fully made" is the phrase used
because everywhere some adjustments have been made. If we have
automobiles and airplanes, we must have systems of traffic
control. Control, by definition, limits freedom; if instrumental
and effective, it confers the more meaningful freedom to go
safely from one place to another. We have begun to clean our air
and make other adjustments, but the process is incomplete. The
villager had an indoor pollution problem that was solvable and
solved by technology. We have pollution problems solvable by
technology also, but they require institutional adjustment.

Being humans gifted with imagination, we can live in
different worlds of illusion. Some can close a doomsday book,
reset the thermostat, and turn off the stereo and the lights,
turn on the electric blanket, and go to bed and have nightmares
about technology out of control. Others can live in a world of
interdependence and imagine themselves as discrete, autonomous,
atomistic individuals. The power conferred by technology fosters
this illusion. In a sense, our technology has not led to too
much centralization but to too much decentralization.

If we are to organize our communities based on shared

beliefs, then we must distinguish between tastes and technological values. Tastes that have consequences for others
should be subject to valuational tests. It may be a matter of
preference where a person wishes to smoke, but it is not a matter
of taste as to whether those in a shared, closed environment
might be subjected to carcinogenic smoke. Civilized communities
may allow people to commit suicide (most people don't, unless
they do it a little at a time, such as by smoking, or according
to ritual necessity following humiliation), but must we allow
people to do it in motor vehicles and take their fellow citizens
with them?

The institutional adjustments we need to make involve
applying the principles of technology (i.e., technological
valuations) to the control of technology. Traffic control is a
good example. We can define the needs of safe, comfortable
movement of people in vehicles (cars, planes, trains, and so on)
from place to place; then we apply engineering principles of
traffic flow and use technology in the form of hardware to
achieve these results. Where there is a shared understanding of
technological valuations, regulation can work to the benefit of
all. It brings institutions and behavior into adjustment with
the human possibilities of technology.

Regulation can work in other areas. The reduction in the
flammability of pajamas as regulated in the United States has
led to a dramatic decline in children being badly burned or
dying from burns. This decline has led to the actual closing of
some children's burn hospitals. Similarly, putting child-proof
caps on medicines and poisons has led to a dramatic decline in
children being poisoned in the United States. "Accidental
poisonings of children under five have dropped by about 60% since
1970, and deaths from aspirin ingestion alone have been cut by
83%" writes Himowitz (1983).

From 1973 to 1980 this meant the prevention of "an estimated
200,000 accidental poisonings of young children" (Castle 1980).
Clearly, with the help of intelligently conceived regulation and
new technologies, we can continue to make our lives safer and
healthier. Smoke alarms can save lives and property. Food
additives and preservatives may not always be safe, but not
using these substances can be even less safe. Food poisoning in
modern industrial societies is a rarity; it was commonplace in
preindustrial societies. If we are willing to bear the cost, we
can have greater testing of the chemicals we use to further
increase our safety. Whatever the failures of regulation, they
pale compared to earlier conditions when we either did not use or
did not have the technology, or we used it indiscriminately and
unsafely. From about the turn of the century until the mid-1930s
in the United States, the famous "muckraking" literature is
replete with horror stories of medical quackery and contaminated
foods.

The right to choose does not guarantee the right choice.
What may seem today like the wisdom of the ancients is more
likely a bit of sagacity mixed with a lack of knowledge. The
wise could appear to know much about everything because so little

was known. Today the collective scientific and technological
knowledge is vastly beyond the comprehension of any individual.
The very technologies that we exercise operant control over are
technologies of so many different kinds that we may know how to
make it work (that is, flip a switch) but not know how it works.
We consume so many other things, from food to music to medicine,
where our individual fundamental knowledge is limited.

Lest we seem to be supporting the runaway technology thesis,
let us note that this has always been the case. Even with their
more limited diets, our ancestors probably knew less about the
nutritional quality of their foodstuffs than we do. The farmer,
too, of centuries back, or the metal worker, may have known
intimately the processes in which they engaged, but that does not
mean that they understood agronomy or metallurgy as we do today.
The problem today is that the gap between individual knowledge
and collective knowledge is great and growing. The cure for
alienation, if that is what we suffer, is more knowledge, not
more ignorance. The disparity becomes critical when what we do
not know is needed for our more effective use of technology.

The intentional withholding of this knowledge gives others
power over our lives and limits the decentralizing potential
of technology. Just as technology is not fully transferred
from country to country or from culture to culture unless there
is also a knowledge transfer, the power over technology by
the consumer within a country or culture is limited unless
there is access to some minimal knowledge about it and/or to
some collective means of more complete knowledge. That is why
the regulatory process in many countries involves both the
restriction of the deleterious (for instance, flammable pajamas)
and the required provision of information, however minimal, that
facilitate intelligent choice. Technology's purposes are human
purposes, and regulation can be oriented to the fulfillment of
these objectives. Technology use within a society is incomplete
(out of control for the individual) without some knowledge basis.
Here also, intelligent regulations beneficial to the community
are derived from technological valuations.

James Grant (1984, 45), writing for UNICEF on primary health
care for children, spoke of strategies that "empower" people. If
a technique such as oral rehydration therapy (ORT) can cure the
dehydration associated with diarrhea, then it truly does empower
people. It empowers them in the most fundamental way; namely, it
prevents the deaths of children, or stated differently, it allows
the life process to continue. The fact that in villages throughout this globe the basic means for the therapy are available
and comprehensible is certainly an added benefit. As we have
stressed, technology transfer is incomplete until one is able to
make it one's own. However, Grant speaks of other technologies
as serving to alienate. The concept of empowering people
is frequently used by appropriate technology advocates; but
what about immunization? Here we have vaccines developed by
scientists unknown to the beneficiaries and frequently delivered
with sophisticated equipment on a mass basis. Immunization
empowers people because it is a cheap, effective way to solve

human disease problems. Nor is ORT alienating because it was a by-product of research on cholera, and the technique is promoted by a variety of sophisticated marketing strategies using technologies such as radio.

The very characteristics of the technologies touted as being participatory are found increasingly in the technologically advanced societies. In medicine, increased knowledge of health and factors contributing to longevity have brought about changes in peoples' lifestyles. People have been empowered (self-imposed) to live longer by their own actions of exercise, not smoking, and eating certain foods and not others. Nevertheless, when the need is for other types of medical intervention, very few who can afford it, forgo it. Most people would like "to become informed activists in the protection of their own and their family's well-being." How many proponents of participation, though, would forgo necessary lifesaving surgery because it made them "passive and dependent recipients of health care?" (Grant 1984, 45). The consumer movements are an attempt to provide expert judgment to the recipient of their product or services. To the extent that these movements and other sources close a portion of the information gap, then people are empowered to make more intelligent choices. Other devices, such as second medical opinions prior to surgery. also contribute to this process.

The critics of modern technology have falsely claimed that mass technology has dehumanized us because it requires fewer skills to operate. The alienation or passive dependency arguments are based on the contrary assumption: namely, that the benefits of modern technology are derived from such a vast sum of knowledge and skill capability that any of us or any small group of us have only the smallest piece of it. In the late 19th and early 20th century the idealistic belief existed that the rise of professionalism would provide the standards of conduct from which we would all benefit. Without question, this vision was noble; it was also utopian. The rise of professionalism and professional standards has not been the unvarnished blessing to which men of good will aspired (Haskell 1984).

We need not despair the failure of professionalism. If one looks at professions today and a century ago, it is hard to argue against the proposition that standards have been set and raised and the public has benefited. In a sense, professionalism is the application of technological values in a particular area of endeavor. Unfortunately, the process of setting standards confers power upon those who control entry and exit. The same system of standards that keeps out quacks and incompetents can keep others out, for reasons of prejudice, ideology, or simply the economic gain of those already in.

We face a dilemma. As in other technologies, when we create a new organizational technology, professionalism, we may solve some problems but we create others. We can abandon the legal power of certification through governmental licensing and take our chances on a free market (Friedman 1962, 149-60). Or

we can seek simpler technologies that we all understand and
eliminate the dilemmas. Most of us would prefer to be somewhere
between these two extremes where there is still a vast area for
disagreement. In this author's judgment, it is hard to see any
position between the extremes that does not involve some shared
community process of technological valuations. Essentially we
will have the trial-and-error process of evolving technological,
professional standards and seeking always to minimize the number
of potential abusers.

Most of us are so dependent on the technological valuations
of professionals that we take them for granted. Unfortunately,
at times we accept them as authority rather than as operational
assessments based upon a body of knowledge (Haskell, 1984).
Taking experts as authority figures invests them with non-
technological characteristics bordering on infallibility. When
experts fail, then, it is not a failure of expertise. Science
and technology, like democracy, are self-correcting systems.
Humans are fallible, both those who are elected and those of us
who elect them. If we err on one round, there is always another
round for correcting, unless we make a catastrophic error such as
nuclear war that is essentially irreversible. The individual
tends to be a more likely victim of irreversable error than the
society.

In science and technology the processes of experimentation
allow the testing and retesting of ideas. Falsification is as
essential to the process as verification; they are two aspects
or possible results of the same process. Only through full
disclosure of procedures, enabling others to replicate and
test processes and results, does individual inquiry become
part of that shared public enterprise of science and technology.
The open inquiry central to those processes are part of that
constellation of values we call freedom.

Our concern for safety in modern technology is frequently
focused on cancer-causing chemicals and radiation. We tend not
to notice the reliability and safety that we get from most
things mechanical or structural. Evolving professionalism in
engineering and scientific methods of testing reliability have
given us safety margins and fail-safe redundancies in such items
as airplanes and bridges (Wenk 1979, 165-67). With increasing
legal liability for professional conduct (or more precisely,
professional failure), we are likely to see increasing margins of
safety in most technological endeavors. While the public may
approve, many economists think that we may be buying more safety
than is warranted by the cost.

It has become a cliche that the technologies of communication and transportation have been "shrinking" our world. This,
plus satellites and photographs from earth's orbit or from the
moon, make us increasingly aware that we share one planet. Such
technologies have helped to shape our concept of humanity.
Haskell argues that ethics become operative when the possibility
of action emerges. The market, by interrelating people over a
wide area, is responsible for the origins of what he calls

"humanitarian sensibility." From Adam Smith onward, the market
has been viewed as a system of exchange where all benefit. The
very title of Smith's masterpiece, An Inquiry into the Nature and
Causes of the Wealth of Nations, implies an international system
of mutual benefits. To critics this has been more an ideology
that justifies markets rather than a reality of them. In either
case, mutual benefit has become a standard by which we are
supposed to judge economic systems and transactions. The
broadening of the market by technology, furthermore, has made
this code of conduct nearly universal. Before Adam Smith, the
mercantalistic idea of international exchange was to benefit a
country at the expense of its neighbor. When might makes right,
there is no room for reasoned argument.

However we may fall short of achieving the ideal of economic
interactions that benefit all who participate, it has become an
idea that forms the basic discourse on economic systems, trade,
development, and choice of technology. We do not argue as to
whether technology should benefit people but over which ones
provide more benefit and which ones are possibly harmful. Adam
Smith may have given us the idea that the means of production
and the economy exist to serve people by producing goods for
them to consume. For Smith, and for David Hume before him,
Hume, it was not only a novel but a heretical idea as well that
economic progress could benefit everyone, both within an economy
and external to it. The ideal may have come from the moral
philosophers and economists, but the potential for achieving this
idea has come from the unprecedented affluence created by
technology. Technological values are involved in both the idea
of production for human purposes and the possibility of its
fulfillment.

Technology is a complement to, but by no means a substitute
for, political freedom. Repressive regimes often try to use
the freedom that advanced technologies confer as a means of
buying support for their politically restrictive policies.
Most citizens of industrial countries take their technological
freedoms for granted. Few in a country such as the United States
have experienced the time when electricity did not light up
their home (except on those rare occasions when the system
failed). It is hard for them to understand the excitement of
rural electrification in the third world and the development
possibilities thereby imparted. The enterprise that electrification turns loose, in the form of people acquiring small pumps,
mills, crushers, or other machines, is decentralizing in the best
sense of that term. That the electricity is generated from a
large-scale hydroelectric power project or a nuclear power plant
is of less concern than the fact that it is there. Once having
experienced this, it is difficult to take seriously the silliness
of those who wish to liberate people from the leviathan of
centralized power generation with less reliable local sources.

Some of us have become so accustomed to advanced technology
in our lives that it becomes functionally invisible to us. We
take it for granted until it fails, and then we blame technology,

unaware as to its greater reliability than alternative technologies.   Those of us among the affluent who must fight food
because of weight problems have trouble understanding those who
must fight for it.   The freedom from starvation or even from the
fear of it is a freedom that modern agronomy has given to an
increasing proportion of the world's population.   Beyond the
basics, technology gives us the freedom to do things that we had
not thought of doing before.

The yearning expressed by many for our collective return to
the village is institutional adjustment in the wrong direction.
A return to the village either forces us to use technologies
that drastically reduce our control over our destiny (and the
number of us that the earth can support).   If we tried to use
the same technology as now, only within village-sized units,
we would be assured of the out-of-control technology that so
many fear.   What, furthermore, is to prevent a tyrant from
arising in a village and gaining control over technology and over
an everincreasing number of villages?   "Thinking globally; acting
locally" is a catchy slogan (Feather 1980, 436).   If we are to
take it literally, it implies that thought is somehow divorced
from action.   If we need to think globally, it is because we need
to act globally.   This does not preclude the equal necessity of
thinking and acting locally.

This chapter and others have spoken of the universality
of technology or of technological values.   In using the term
universal the less-than-literal meaning of global has been
followed.   Yet, as our technology carries us out beyond the
planet, we might be wise to explore the possibility of values
beyond ourselves.   Our 4-to-5-billion-year-old solar system is
relatively young in a cosmos that is 10 billion to 20 billion
years old.   Among the older stars there are undoubtedly planetary
systems.   How many times has life emerged and then faltered or
failed, either early in the process or at a point beyond our own?
In all these circumstances, some may have learned to stay the
course.   If we have come so far, so quickly, how far along must
older life forms be?

In the ongoing argument over whether intelligent life exists
elsewhere in the universe, there is a powerful argument against
it.   Von Neumann (1966) and Poundstone (1984) have demonstrated
the possibility of creating a self-replicating machine.   Such a
machine would partake of many of the characteristics of life
itself.   Life forms technologically beyond our own would have
built such machines long ago and programmed them to search the
universe for other life forms.   These machines would go out and
land on planets, then make replicas of themselves which would
search out other planets.   By this method we would have been
found long ago.   Thus, it is concluded, since we have not been
found, there is a strong case against intelligent life elsewhere.
Or is there?

An equally persuasive argument says that perhaps we have
been found and are being watched.   Intelligent life forms that
have explored the universe have undoubtedly seen other life forms

begin and fail. Some may have succeeded in every way but in the
avoidance of self-destruction. We are harmless to external life
as long as we remain planetary. The thesis has been suggested
that we are silently on trial. If we prove to be capable of
controlling ourselves as users of our technology, then we may be
allowed to reach beyond our planet without a threat to others.
If not, what choice would a peaceful life in the universe have
but to eliminate us?

This thesis raises some interesting value speculations.
Life forms that have survived would have the technological values
of problem solving and survival. If they have thus far spared
us, they must have some sense that the cosmos is not their
exclusive preserve and that there is some value in diverse life
forms. Quite possibly, then, they might be judging us on whether
we evidence similar values by seeking to share our planet with
other life forms or with other members of our own species.

In trying to project how other life forms might perceive us,
it is difficult if not impossible not to anthropomorphize their
perceptions. Anthropocentric dilemmas are not limited to the
study of life and society. Cosmologists also speak of an
anthropic principle. The values favoring survival might be
true universals, but, if so, it is true by definition. Is it
scientific speculation or science fiction? If we fail to pass
the test, either an externally imposed one or the result of the
dynamics of our technology and institutions, then the technology
that brought us this far and the values it implied would have
been for naught. We would be a brief flicker in the cosmos where
life tried and failed. In either case it is useful to attempt to
strive beyond ourselves and to look forward. Whether looking at
ourselves from the outside or the inside, the developmental
problem-solving survival values of technology itself provide the
framework for the much sought after intelligent control and use
of technology.

CHAPTER 12
Technological Choice and Choice of Technology

The economic issues involved in the selection of technology were once purely abstract and abstruse. Theoretical exercises involving equations and graphs could be carried out without the merest reference to actual technologies or to any noneconomic variable. The discourse on the issues of technology for development revolved around capital/labor relationships, relative scarcities, comparative advantages, and other questions of factor endowment. One can still engage in such inquiry in economics, but in the emerging development literature the choice of technology is expected to involve a host of concerns, from social policy to the environment. The question of capital intensity is still vital, but it is one aspect of a multidimensional problem.

The awareness of the broader implications of the choice of technology in economic development emerged in the late 1960s and early 1970s. A variety of movements took hold with a common theme that involved the selection and use of technology. Central to all of them was the criticism of modern technology in general and the thrust of technology and technology transfer to the third world since 1950. The critics of the trends of contemporary technology were and remain so strong in their views that they are accused of being antitechnology. Most vehemently deny these assertions. In the developed countries there are the environmentalists, the catastrophists, and the limits-to-growth theorists. For developing countries there are a variety of movements that use defining concepts, such as appropriate technology, intermediate technology, and alternate technology. Of course, it is possible, particularly given the idea of differing factor endowments, to favor capital-intensive technology for the economically advanced countries, and labor-intensive technology for the less developed world. Apart from this latter position, there are some assumptions common to all of them:

1. The exhaustion of important resources in the not-too-distant future, i.e., 10 to 50 years. This would include not only mineral resources but even more important, energy resources.
2. The destruction of the environment by technology in the

163

form of air and water pollution, desertification, and the rapid loss of plant and animal species (particularly in the tropical rain forest).

3. The increasing inequality brought about by the use of modern technology in third world countries.

4. The loss of a sense of community and family and the alienation of the individual as a result of the continued spread and intensification of modern technology.

5. Finally, the redemptive possibilities of reversing these trends through the use of small-scale technologies that utilize renewable resources are community based and controlled and are oriented towards local needs.

SMALL IS BEAUTIFUL?

In many places the selection of energy distinguishes the proponents and opponents of "soft" technology. To some, solar energy is an embodiment of all the small-is-beautiful virtues and nuclear power, the encapsulation of all evil. They have even defined the era we are entering as the Solar Age. In the United States and possibly elsewhere, an equally enthusiastic, ideologically rigid group forms a mirror image to the solar devotees. To this opposing group, solar energy is utopian (or worse, un-American), as is conservation, and our only salvation lies in nuclear power and other high-tech solutions in the marketplace. For these latter groups, having solar power as an energy source for the operation of satellites is an exception to this principle.

Much of the tone for ideological debate on energy policy was set by Amory Lovins ([1976] 1977, 25-60) in his famous article in Foreign Affairs. Lovins established the dualism between hard and soft energy paths. The soft energy path is the one that is morally virtuous and sustainable. The hard energy path is the high-tech, nonrenewable strategy. It is referred to as vulnerable and brittle.

"Soft" technologies are "flexible, resistant, sustainable, and benign." Further, "the distinction between hard and soft energy paths rests not on how much energy is used, but on the technical and sociopolitical structure of the energy system, thus focusing our attention on consequent and crucial political differences." The 5 characteristics of soft energy technology are (1) renewable based on energy income, not on depletable energy capital, (2) diverse, (3) flexible and relatively low level, and (4) and (5) matched in scale, geographic distribution, and quality to end use needs.

Current energy technologies have two uses, according to Lovins. First, there are the technical fixes. Technical fixes are a variety of technologies, such as insulation, that can "plug leaks and use thriftier technologies" without our having to change our lifestyle. Second, current technologies, properly used (such as technical fixes) allow us to make the transition to the soft technologies that are right, both in terms of hardware and social change. If we use these technologies for any purposes other than transition to soft energy, then we are burning the "fossil fuel bridge." The key to Lovins' thesis and to that of a large number of the alternate technology proponents is that the "two directions of development are <u>mutually exclusive</u>" (emphasis added).

Not all proponents of appropriate technology are committed to it to the exclusion of all other technologies. There are intelligent, pragmatic individuals working throughout the globe, using whatever technologies are available, from satellites to small cook stoves, to solve problems of development. Many of these people find appropriate technologies to be the type most suited to the problems with which they are concerned and feel no need to denigrate other types of technology. This group should be distinguished from the larger body of authors (only some of whom attempt to be development practitioners) who have an apocalyptic view of modern technology and a purist, utopian vision of small technology.

No discussion of small technology is complete without mention of E. F. Schumacher. His treatise, <u>Small Is Beautiful</u> (Schumacher 1973), has become almost holy writ for many of his followers. For Schumacher, our choices are even more constrained than they are for Lovins, for he maintains modern technology may not be around much longer for us to mistakenly choose it: "In the subtle system of nature, technology, and in particular the super technology of the modern world, acts like a foreign body, and there are now numerous signs of rejection" (p. 139). Schumacher wishes to replace mass production with production by the masses: "The technology of <u>mass production</u> is inherently violent, ecologically damaging, self-defeating in terms of non-renewable resources, and stultifying for the human person." In contrast to mass technology, <u>production by the masses</u> uses "the best of modern technology and experience, is conducive to decentralization, compatible with the laws of ecology, gentle in its use of resources, and designed to serve the human person instead of making him the servant of machines" (p. 145). Schumacher calls his kind <u>intermediate technology</u> and indicates that it could also be called self-help technology, democratic technology, people's technology, or technology with a human face. With all these wondrous characteristics there is only one small problem, that the technology does not exist or did not exist when the book was written. "Although we are in possession of all requisite knowledge," he writes, "it still requires a systematic, creative effort to bring this technology into active existence and make it generally visible and available" (pp. 145, 146).

If intermediate technology did not exist a decade or so ago, there is no certainty that it exists yet today. It is, to say the least, <u>a prioristic</u> to make sweeping claims about a technology that was not yet created, applied, and tested. Those who are proposing exclusive reliance upon a particular type of technology are in effect advocating a revolution in technology and human society more complete than any we have previously experienced.

Previous chapters have talked about dramatic, revolutionary changes in technology. Throughout these transformations, earlier technologies have continued to survive. It is true that some of the earlier materials used in tool-creating, such as flint or poorly worked metals, are essentially historical artifacts. However, some of the basic tools that are still used in home and industry have their origins (in cruder forms) among our earliest tools. Technologies co-exist. The fundamental reason is that technology is a problem-solving process. A new, "superior" technology may supercede an earlier technology for a wide range of problems but not necessarily for all earlier technology uses. A hammer, however primitive, is the most cost-effective, convenient tool to pound a few nails in around the house.

Here again the technology of printing can serve as an example. Printing by movable type was eventually to bring about a qualitative and massive, quantitative change in book production and a revolution in knowledge. From early printing to printing by movable type there were a number of changes. Initially, the new form was used for the luxury trade, though its most important contribution was to be in mass printing. In its infancy, printing was inferior to other forms of reproduction (Usher 1959, 254-55). In 500 years we have gone from printing to a vast array of types and devices for imprinting the work and reproducing it. Printing replaced handwriting for reproduction but did not replace handwriting entirely. The typewriter and other devices replaced handwriting for many purposes, but not all. Handwriting and instruments used for handwriting survive even though they were replaced for many uses. The instruments for handwriting, moreover, have continued to experience technological change. Technologies not only coexist, but they can be interrelated. These words initially were written by pen, then typed, then entered into a word processor, and, finally, reproduced photomechanically.

Problem specification becomes central to understanding the continued coexistence of technologies and to selecting technologies for current needs. If terms such as appropriate technology are to have any substantive meaning, then their appropriateness must be defined in terms of specified problems and specified conditions that form the context in which the problems will be solved.

Placing several a prioristic constraints on the types of technology that can be used to solve today's and tomorrow's problems reduces significantly the probability of an appropriate solution. It is ironic that in Soft Energy Paths: Toward a Durable Peace, Amory Lovins (1977) titles one of his chapters "Technology is the Answer: But What Was the Question?" As this book has tried to understand them, science and technology are both an accumulation of means and methods for solving problems. If we mean by technology the sum total of our problem-solving knowledge at the time and place we seek to address the question, then technology is the answer to energy questions today and in the future. A soft energy path is an answer for tomorrow's energy question before we have an understanding of tomorrow's technological means.

Lovins is correct in that the choices we make today in energy or any other endeavor influence and constrain the choices we can make tomorrow and the day after. This is true both in the cumulative nature of technology as we ease it in one direction or another and in the cumulative nature of societal and economic processes, as we build structures, and acquire knowledge, skills, vested interests, and emotional attachments. Our future possibilities are shaped by today's decisions. Many specific proposals and ideas of Lovins are imaginative and worthy of being weighed on their individual merit, apart from his broader advocacy. However, valid though some of his countentions are, they do not necessarily support his thesis. The cumulative nature of change cuts both ways. If the overall direction of change in which Lovins wishes us to move is in error, then in fact we have made more difficult the necessary technological transitions.

Part of the confusion in the alternate technology conceptualization concerns the sustainability and renewability of resources. If mineral, land, and energy (except solar) resources are fixed and finite, then sustainability is exclusively a matter of conservation. Under these circumstances, entropy means a continuous decline in resource availability. Given these assumptions, the concept of entropy is correct as used by alternate technology advocates, for what we call resources are concentrations of minerals or energy or vital elements in soil structure. No matter how efficiently we build an engine, the energy leaving it is more diffuse and less usable than that which entered it. No matter how much conservation we practice, soils are lost and minerals are leached or used by plant and animal life. There are also reconcentration processes at work that are both natural and human. The human processes of reconcentration require energy.

The essential difference between a renewable and non-renewable resource is the time ratio between natural reconcentration forces and human use and dispersion. When slash-and-burn agriculture is practiced by small populations, the soil is a renewable resource. Long fallow periods allowed the invading vegetation growth to reconcentrate in the soil what agriculture had withdrawn. More continuous agriculture requires a more direct human reconcentration of soil constituents by manuring and/or fertilization. As modern agriculture shows, we can not only reconcentrate, but we can enhance and create by adding in what was not previously there.

Geological forces have concentrated in different locations mineral substances that our technology allows us to use. These geological processes have operated historically and are operating today (although there are some differences, compared to when the earth was very young). For some minerals, such as iron and aluminum, current technology allows us to use economically the concentrations that are only about 4 to 6 times greater than crustal abundance. Random factors, plus such processes as leaching tropical soils of minerals other than aluminum, make it likely that we will be able to use these mineral concentrations for some time to come. At the other extreme are minerals such as lead, gold, and mercury. Current use of these is about 4,000 to 25,000 times (for mercury) the concentration for the earth's crust. These are likely to remain scarce. Most of the minerals that we use and process from ores concentrated 100 to 1,500 times above crustal abundance. The geological processes of concentration are vastly slower than the human processes of dispersion, so we call all mineral resources nonrenewable (Braun 1984, 79).

Solar energy is diffuse as it lands on the globe, and it is because of this low concentration that many see direct solar energy as uneconomical. Natural forces do concentrate and/or store this solar energy. Plants take it in and store it. Wind is a potential source of energy. Water evaporated takes in energy. If it falls on land above sea level, then the gathering stream or river as it moves towards the ocean is a concentration of energy. Presumably, then, if we use these energy sources within their budgetary limits, then we are living off our energy income and not off accumulated and depleted energy capital. If one of the uses of this energy is to reconcentrate the proper constituents of the soil, preferably with renewable energy aggregates such as green manure, then here, too, we would be within the budget. The sun as an external source of energy does allow the alternate energy advocates to relax some of the constraints of entropy, but not all of them.

Once the entropy constraints are allowed to be relaxed because of a sustained external energy source, then we relax or challenge the other assumptions of the alternate technology movements. Energy and technology allow us to take ores and concentrate the minerals we need. Energy allows us not only to reconcentrate but, in fact, to enrich the soils. Changing technologies allow us to do these things ever more efficiently. Utilizing these energy sources, be they solar or nuclear, previously accumulated or on current account, is a matter of technology. If we are looking to sustainability, then we need to turn from the idea of fixed finite resources. Sustainability here is merely one of postponing doomsday. What we need, though, is the sustainability of the resource-creating processes of science and technology. It is precisely these resource-creating processes that are threatened by restrictive technology processes.

Historically, a technology has given us the power to use a resource it created, be it energy, mineral, or land, to sustain ourselves and to sustain the technological process. At any given point in the process, current technology is using resources. Every technology, then, is in a sense a transition to the next technology, which creates new resources to be used in transition to the next technology. Policy diversions, such as an exclusive reliance on the wrong technological path, can make the transition more difficult. This is Lovins' argument applied to his own energy strategy. As to Schumacher, the idea that we can expel modern technology as an alien body (whatever that could possibly mean) but still have all of the knowledge to create his intermediate technology is to posit an argument without any historical or contemporary substance to support it. The only technology that is gentle in its use of resources is one that creates resources.

The catastrophist has given us the image of the overcrowded lifeboat with more people in the water seeking to get in. If the boat sinks, all are lost. This metaphor has a meaning that has a twist unintended by those who use it. For an overcrowded lifeboat is a technology that is inadequate to the task: it is a technology that is too small. It may well be that there needs to be a limit to population growth, but if we are to sustain those in the lifeboat (or planet Earth), then we need boats (technologies) with all the requisite capabilities of doing the job.

In the choice of technology, the issue of moral and social dimensions has rightly been raised. The lifeboat ethic is again instructive. Inadequate technologies can force upon us undesirable, if not morally repugnant, decisions. Either we let everyone on the boat and it sinks with all lost, or we only allow those not on it to drown. Having better technologies does not make us more moral, but it does give us moral options. A larger lifeboat with room for all gives us moral options that are ethically and aesthetically more pleasing. This is more than just a metaphor; it is a reality. In very poor countries where the births of siblings are spaced close together, custom dictates which child is more likely to die. If the first child is immediately weaned when the second child is born, then the first child's probability of death rises significantly. If the culture is one in which both children are given access to the mother's limited supply of milk, then the older, larger child is more likely to survive than the younger. Consciously or unconsciously, the parents or the culture make survival choices for children that others, with access to greater nutrition for the mother and children and access to means of better spacing of children, do not have to make. For those with adequate technology the lifeboat dilemmas, except for some rare or freakish circumstance, are strictly for discussion in ethics or religious classes or for phobias in doomsday books. However masked, the children's death illustration is but one example of painful choices imposed upon people by lack of technology.

The argument against alternate technology theories is not an argument against technologies they advocate. Opposition to a strategy that places exclusive reliance on solar power (in its various forms) is not an argument against solar power.

TECHNOLOGY IN CHINA

There is some experience on the types of technologies and technological strategies advocated by Schumacher and Lovins.

Smil (1979b, 17) writes of "a recent voguish trend in the Western academic writings" about what was "China's everyday practice for two decades." A cursory look at China's efforts in using an intermediate type of technology can be instructive both as to its potential benefits and its limitations. Advocates of alternative technologies have frequently used China as an example of the success of small technology despite the fact that the governing philosophy of China was expressed in the phrase, "walking on two legs." Not mentioned is the estimate that "a third of China's 7 million or so small biogas digesters are under repair" (Hall 1984, 42). This meant developing both large complex enterprises as well as simple projects in rural areas (Smil 1979a, 279).

In developing its energy sources China used a variety of different technologies. Though not the first to use biogas (it was developed much earlier in India), China has used it more extensively than any other country. China opened small coal mines and built many small earthen and rock dams for hydropower as well as for irrigation and flood control. Smil concludes that the "advantages of intermediate energy technologies for China are indisputable."

China is also known for its use of renewable resources and small technologies in other areas. In agriculture the stories are legion about the farmers gathering the night soil from the towns and hauling it back to the communes. In addition, China also developed small-scale fertilizer plants to supplement the use of manure. The electricity generated by the small hydroplants was used to power small rural industries. Many of these industries turned out cement used to line irrigation canals and small farm implements for agricultural mechanization. In all these areas the technologies fulfilled the criteria established for appropriate technology. In addition to being small, they used local resources, frequently renewable, and they used local labor and skills and were generally more labor-intensive than comparable technologies in developed countries.

However useful and necessary these technologies were and possibly remain for Chinese economic development, they do have their limitations. For example, the small coal mines were not always economical. "The lifetime of many of the small mines was ephemeral," writes Smil (1979a), "and a large part of production was consumed in an equally ephemeral iron-making campaign. In fact, it appears that in many cases the human and animal energy necessary just to open, operate, and maintain the small mines surpassed actual energy yield." China has been moving toward opening larger, more capital intensive mines (Rawski 1980, 94).

Along with the small dams, China has constructed some large ones. The total system of large and small dams has increased irrigation and electric power. The flood control system involves the dams, weather satellites, and a rainfall monitoring system tied together with a computer in Beijing (Oka 1981, 134-36; Parks 1982). Given the history and frequency of flooding on the Chinese rivers, particularly the Yellow River, and the enormous toll in lives that these floods have taken, even if some of the claims for flood control are exaggerated, this still is an extraordinary achievement. The construction of large dams has led to the destruction of small ones. Large dams are generally more efficient in electricity generation and are more secure in storing water against the threat of drought. Moreover, they are much more effective against the raging torrents of floods. As with many modern technologies, large dams substitute reliability for the redundancy of many small dams.

China has a long history of using manure in agriculture. Current use is very costly in terms of labor. Overall, at least 20% of total agricultural labor time (and often 30-50%) was absorbed in collecting, processing, transporting, and applying organic fertilizers, which made their use both inconvenient and expensive in areas where there was a seasonal labor shortage (Weins 1978, 682). Citing a Chinese economist, Rawski (1979, 95-96) uses the figure of 30-40% of the "total amount of manpower and animal power expended in the whole year." Furthermore, contrary to popular belief in many areas, organic fertilizers are less efficient as nutrient sources than are the manufactured kind (Rawski, ch. 7). "With organic fertilizers, it is difficult to obtain nutrient proportions that match plant requirements," writes Weins (1978, 682). "Since release rates are slow, it is impossible to insure concentrated release at appropriate periods of plant growth." Organic fertilizers are most effective with use in a regimen that includes chemical fertilizers. Use of organic fertilizers alone traps the user in a "vicious circle" that limits further development.

China has received high praise for its innovative methods for small-scale manufacturing and the use of chemical fertilizers. These small factories produce either ammonium bicarbonate or aqua ammonia, which have a nitrogen content of 16-17% and 14%, respectively. This nitrogen content is much lower than that for ammonia urea. These fertilizers deteriorate rapidly and are difficult to transport (Etienne [1977] 1978, 425-40). It is understandable, then, that since 1972 China has been a major buyer of large, modern, ammonia-urea fertilizer plants.

One of the tenets of the appropriate technology movements is the need for technologies that use surplus labor or save on scarce capital. According to some observers, rural industries in China did not use a large amount of labor. The core of China's industrial development, however, was and remains large-scale industry. "The small direct labor-absorbing effect of rural industries is regarded in China as an accomplishment rather than a failure," writes Riskin (1979, 60, 69), yet "intermediate technology has always been considered a supplement to, not a substitute for, this sector. . . . An essential corollary of this rationale is, of course, that the intermediate sector does not further constrain modern industry by competing with it for factors of production."

Even with the many advantages of small rural industries, large industries have the potential for greater productivity. Rawski (p. 94) found that "despite its capital intensity, Western

industrial technology is so productive that it often maintains
an advantage in cost as well as quality over the alternative
methods, even when, as in China, the ratio of capital to labor
costs is enormously higher than it is in countries for which the
technology was originally intended." The trend in China has been
"consistently toward substitution of capital for labor" (p. 44).
In some instances, capital investment per unit of output was
greater in relatively small plants.   Whatever the benefits of
China's rural industries, "the capital-saving contribution of
small plants is of little significance" (p. 48).

     The small hydroplants and the rural industries provided
the energy and the implements for irrigation and agricultural
mechanization.   In many circles mechanization is seen as an
evil, since it displaces labor.   This is an ongoing argument
in development, as many counter that even in areas of labor
surpluses there are peak periods in which labor is in short
supply.   Further, with the development of miracle grains,
multiple cropping, and fertilization, there are more peak demand
periods and more problems requiring mechanization. Studies for a
number of areas, such as the Punjab in India, have found that
mechanization, used with these other technologies, can actually
increase the demand for labor.   For China, even Mao recognized
that mechanization is "the fundamental way out for agriculture"
(Maxwell 1979b, 87).   Not only is labor a constraint today, but
"even in the 1930s, most areas of China experienced labour
shortages in one or more peak agricultural seasons during the
year" (Riskin 1979, 60).

     The small technologies used by China illustrate their
inherent limitations.   The assumption of finite resources is in
fact self-fulfilling, and at a lower level than realized.   For
example, on the use of manure as fertilizer, there are two
constraints that put a ceiling on expanded utilization.   First,
there would have to be more humans and/or animals to provide the
manure.   As China's current policy shows, more mouths to feed is
not the way to get more food per capita.   More animal manure
would require dedicating more land to raising stock, and this
would likely be counterproductive to growing more food.   Second,
given the large expenditure of human time and energy to gather
and use the manure, there is a severe limit to how much more
manure could be used, even if available.   Of course, some of that
constraint could be broken with mechanization, but that leads to
the path of "nonrenewable" resources.

     There is even a more egregious example of the limits to
labor-intensive technologies.   The Chinese have straightened
river beds to gain more land.   In one instance they have actually
tunneled underground so that the river bed could be added
to land for agriculture (Fairbank 1979, 3-4; Maxwell 1979b,
65-66).   Once a river is straight, no more land can be gained by
surface manipulation.   Once a river is underground, further
tunneling is superfluous for gaining (or "creating," to use our
earlier terminology) land.   China has made great strides with
labor-intensive efforts, which have allowed them to stabilize

agricultural output and absorb and feed a vast increase in population. Too much of this has been achieved by increasing the number of workdays per year. The absolute limits to this process are the number of days in a year. The practical limits are much less than that and have probably been reached. Further, the purpose of development is not to make people work more but to allow them to work less and produce more. The way this is achieved is through the open-ended path of technological change to which China is committed.

Vaclav Smil (1979b, 17-18) sums up the Chinese experience with small technology in a memorable phrase: "The Chinese, the world's greatest practitioners of frugal smallness, discovered that small is useful but small is not enough."

STRATEGIES FOR THE USE OF TECHNOLOGIES

Some modern technologies possess some of the characteristics ascribed to intermediate technologies. The miracle grains have proved to be scale neutral. D. Gale Johnson (1981, 291) asserts that small farms have been quite capable of adopting the rather complicated technologies that were involved. The amount of labor per unit of output fell (which is good), but the increase in output was sufficiently large so that farm labor employment grew. Moreover, as Johnson and others have shown, the poor were primary beneficiaries. These grains were almost rejected by India in the mid-1960s out of fears similar to those raised by the small-is-beautiful proponents. Without doubt, the arguments about scale of technology are more than merely academic. If Johnson and this author are correct, successful opposition to new technologies, such as those of the high-yielding varieties of grains, could have had extremely dire consequences for tens of millions of the world's poorest people. If we are wrong and the small-is-beautiful proponents are right, then large-scale technology is leading us to global destruction. Issues of economic and technological policy do make a difference.

Keith Griffin (1972, 62, 64) provided one of the most balanced assessments of the green revolution. As with other authors, he found that "the majority of technical changes that are associated with the 'green revolution' are neutral to scale and therefore do not give the large land owner a cost advantage over the small peasant." Though the technologies may be inherently scale neutral, they are used in the context of social and political structures that may deny equal access to them (Goodell 1984). Farmers without credit for seeds or fertilizer or access to water cannot compete, not because of the technology but because it is effectively denied to them. As Griffin observes, the problem results "from the bias of government policy and the fact that public institutions are not scale neutral."

Given the non-neutrality of institutions, a donor concerned with helping the very poorest might select a soft technology because they lacked the control over the institutional forces that led to greater inequality with other technologies. A

government having the same objectives might opt for a more
advanced technology and social reform. Or a donor might provide
the key inputs--credit, extension, etc.--that make for a wider
distribution of the benefits of a technology that would produce
greater inequality under the existing institutional structures.
Or a government might find it easier to promote soft technologies
rather than fight entrenched interests and their control of vital
factor inputs. Small or soft technologies can be a vital part of
any development strategy for many different kinds of reasons, but
over the long haul, these technologies cannot be a substitute for
genuine institutional reform.

A serious problem with soft energy technologies is that they
do not always fit the circumstances of the environment or their
end use. This is ironic, since these are considered unique
strengths of these technologies. It has long been recognized,
for example, that one of the many drawbacks of solar cookers is
that they cannot prepare the main meal of the day at the time
that many eat it, well after dark. Smil (1979b, 12-15) shows
generally that soft technologies frequently fail in end use and
cannot provide energy for two of the most critical needs, the
production of fertilizer and good-quality pig iron. Nor does
solar energy have the ability to provide the energy to create
high-quality silicon necessary for photovoltaic solar systems.
It is therefore not self-sustainable (Beckman 1984, 417). The
sun may shine everywhere, but it does not mean that solar energy
always reaches the surface at effectively usable intensities;
nor is wind regularly and reliably available in some of the
areas, such as India, most in need of energy. Also, for some
of the green biomass energy sources, these materials have
"indispensable" alternative uses, including soil protection.
Smil's argument is that it is the poorer countries in the tropics
that are generally most deficient in these soft energy sources
purported to be the solution to their problems. Whatever
their deficiencies, modern energy technologies tend to be less
constrained by environmental factors.

Quite likely a strategy of using different kinds of technologies may be the optimal one for a developing country.
Technology transfer is, as we have argued, a form of invention.
Every transfer, every project involves some experimentation and
therefore some greater than normal chance of failure. Some argue
that many small projects instead of a few large ones are a form
of redundancy against the adverse effects of failure (Caiden and
Wildavsky 1974, 308-13). One can argue equally well that a
replication of small projects that involve the same type of
contingencies does not necessarily provide safety in numbers. A
number of small hydropower plants can fail as easily as one if
there is a drought. However, differing technologies drawing upon
differing resources and capabilities can provide a buffer against
a variety of adverse circumstances.

Whatever the mix of technologies may be, the totality of
them should be understood, organized, and used as a technological
system. It is important that there is a functioning linkage

between the technologies, for one of the overriding advantages of more advanced technology is that it is the kind in which worldwide research and development is continuous and substantial. The dynamics of technological change is a beneficial element that should be built into every economy and techno-system.

Without these linkages, technologies tend to fail. Too often a particular type of technology, an industrial plant or an agricultural practice, was lifted out of one environment and placed into another, with disastrous consequences. Jane Jacobs, in a perceptive article, shows that large enterprises depend upon "innumerable small firms" that provide them with a variety of specialized technological inputs. The economies of scale are the result of a concentrated mass of enterprises (that is, industrial urban centers) linked together by a market. It is easy to understand how an industrial plant placed in an environment without these services cannot survive. One of the advantages of very large enterprises is that they can establish plants in different regions within a country or in other countries and have a large degree of technological self-sufficiency internal to them (Jacobs 1984, 41-42).

For small developing countries, developing the infra-structure for supporting technologies is a long, difficult task. Even small-scale technologies require a technical support structure for maintenance and replacement. Furthermore, the older, used machinery frequently recommended for developing areas is most in need of this support structure, particularly for spare parts and maintenance. When spare parts are no longer sold, such equipment may be more appropriate to a larger industrial country with machine shops that can manufacture the parts.

Modern transportation and communication can provide a substitute for a geographically concentrated system of tech-nologies. If a crop fails because of late rains, seeds with a shorter growing period can be flown in from a research center. Satellite communications can link people to centers of operational knowledge. Also, if the repair function is important enough, parts, technical personnel, and so forth, can be flown anywhere in a matter of days. We do have a global technological system. Developing countries need to integrate their tech-nologies and see that they are effectively linked to scientific and technological endeavors in agriculture, public health, industry, or whatever. Governments may use satellites for television or radio communications to teach mothers about oral rehydration therapy and how to save their babies from death from dehydration. Satellites can be as much a part of the technology that empowers people as are the locally available ingredients that provide the cure. A farmer working with a variety of modern seeds and traditional crops may draw water from a tubewell or from an irrigation canal from a modern dam project. These technologies can coexist and interact fruitfully. What we are advocating here is what many people in development planning are attempting to do.

There is a widely quoted adage in development, that if you

give a man a fish, you will have to continue to give him a fish, and he becomes dependent upon you. But teach a man to fish, and tomorrow and thereafter he can catch his own fish. Important though that may be, what is described here is technique transfer and not technology transfer. Teaching a person to fish bumps that person up to a higher level of stagnation. It may well be the case that this is all the outside donor or the development planner can achieve. If so, we should clearly recognize the nature of the process and its limitations. As we have argued, many of the small, soft technologies are inherently limited in their development. Their assumptions about resource limitations are self-fulfilling.

The self-limiting character of the soft technologies cannot be stressed too much. The arguments of soft technology advocates about limits, finiteness, and entropy apply most precisely to the technologies that they advocate. There is an interesting analogy (and only an analogy) between the anaerobic fermentation processes of the biomass energy systems and the transition from early procaryotes and other life forms. Day (1984, 88) writes, "Fermentation is a low-yield energy source. And for those organisms which rely upon it as the sole means to fuel their process, they are forever condemned to a low level of existence." With life and with human societies, living within limits is a proscription for stagnation. Success comes from finding means to overcome limitations.

Technology transfer and development involve teaching our person the process of fishing and somehow linking him (or her) to that activity as it is carried on elsewhere. Self-reliance is a worthy goal, but it does not have to mean autarky or local, inward-looking policies of using technology. Self-reliant individuals and groups can also interact and be interdependent. What is important about technological choice for the individual, small groups, countries, and so on, is that it is choice. There are many dimensions to the choice, but essentially people are choosing a technology because it provides more or better (by their values) of something they want. By becoming involved with open-ended technologies that are themselves experiencing ongoing development, it is far more likely that the choice of technology will involve a continuity of desired developmental changes. With this kind of linkage, rather than through passive dependence, people can acquire an enhanced ability and freedom to continue to choose, or reject, technologies.

CHAPTER 13

Technology, Science, and Quality of Life

The critics of modern technologies frequently refer to values or ways of life that we have lost. If modern technology is alienating, it is because it is separating us from something to which we were previously attached or attuned. The idea of an earlier golden age is not unique to contemporary civilizations. In these golden ages, it is generally not difficult to find writers and others who were themselves yearning for the lost virtues, or innocence or greatness of an earlier time. The ages of heroes, or towering giants, or Camelot are either in the past or in some utopian vision of the future. Unfortunately, closer scrutinies of reality have yet to locate a lost paradise on Earth.

Lewis Mumford was one of the first to distinguish between small, democratic technologies and large, centralized totalitarian technologies. He called mining a dehumanizing technology that formed "a concrete model of the conceptual world which was built up by the physicists of the seventeenth century" and of subsequent developments of science generally. Comparing the work of the miner to that of the European medieval farmer, Mumford (1934, 79) found that "no pretty wench is passing in the field with a basket on her head, whose proud breast and flanks remind him of his manhood: no rabbit scurries across his path to arouse the hunter in him: no play of light or a distant river awakens his reverie." Any reading of the history of the period will show that work in agriculture was anything but idyllic.

Though the apocalyptic metaphor is frequently applied to the present and immediate future, it is clear the famed Four Horsemen, war, famine, pestilence, and death, rode far more frequently in the past than they do today. If Mumford's yeoman had been working during the fourteenth century of the Black Death, he might have been one of the millions killed in an epidemic that in some areas claimed half the population. Estimates for the deaths run to 25 million in Europe, 60 million worldwide. Some estimates run to one-third of the world population, which is probably quite a bit on the high side (Cornell 1976, 159–60; Asimov 1979a, 79; Asimov 1979b, 251). A series of recurring plagues that swept Europe and Asia in the previous millenium may have taken a total of over 100 million lives; in more recent times (1918-1919) Spanish influenza killed anywhere between 25 million to 50 million people (Cornell 1976).   One could go on at great length cataloging one after the other, disease epidemics where the cost in lives was hundreds of thousands if not millions.   Our modern systems of public health, immunization, disease control and other forms of intervention make similar occurrences today highly unlikely both in terms of the number killed and definitely in terms of the proposition of the population afflicted.

The Four Horsemen have tended to ride together.   Wars cause disruptions in social structure and in production of essentials such as a food.   Inadequate food, whether the result of war, social disruption, or natural disaster, makes people more susceptible to disease.   Various combinations of these raise the death rate.   Natural disasters have made their contribution to this toll (Cornell 1976).   These, too, form a macabre catalog in human history where the numbers have also run into the millions. As noted previously, the death rate from such occurrences today in the developed countries is about one-tenth that of the less developed world and both are far below the historic norms. Though science and technology cannot prevent those hazards (except floods), it can in the case of severe weather, volcanic eruptions, and, possibly soon, earthquakes, give advanced warning so that precautions or flights are possible.   Better construction gives greater protection against the elements.   The modern economies and technology allow for more rapid and effective post-disaster relief.   For many of the natural disasters--wind, rain, floods, and earthquakes--the preponderance of deaths came after the event from the spread of diseases that resulted from the breakdowns in food, clean water supply, sanitation, and social structures.   The comparatively high figures that prevail in poorer countries (in relation to developed countries and not to past experiences) are now being reduced through satellites and better coordinated systems of weather monitoring and forecasting and international systems of emergency relief aid.

Any coverage of major disasters inevitably discusses floods on the Yellow River of China.   It is estimated that the river has flooded 1,500 times in the 2,000 years before 1949 (Oka 1981). The name "China's Sorrow" seems well deserved.   Estimates for some nineteenth-century floods run into millions killed.   With systems of dams, however, the Yellow River has basically been brought under control.   Under this regime there may in the future be problems, floods, and lives lost, but certainly not anywhere near the level of previous disasters.

The subject of disasters, natural or human created, is incomplete without discussion of famine.   As Amartya Sen (1981, 1-8) has demonstrated, famine is not always caused by natural forces.   He distinguishes between starvation caused by general lack of food and starvation caused by lack of "entitlement" to food.   In the cases he studies, "food availability decline" was neither a sufficient cause to explain famine nor, in some instances, a cause at all.   If food is exported from a famine region or even if food production is up from the previous year, then lack of "direct entitlement to food" must be seen as a primary cause (Sen, 154-66). For these reasons Sen does not find much benefit in trying to understand hunger in the world by relating population to total food supply. Consistent with our earlier analyses, Sen found "as far as the present is concerned—rather than the future—there is no real evidence of food supply falling behind population growth for the world as a whole."

Other than the Bengal Famine of 1943, all of Sen's detailed case studies are post-World War II. Though not always a cause of famine, Sen does recognize that with a continued decline in world per capita food supply, "starvation would be sooner or later accentuated." The same reasoning applies in the other direction, that a lower per capita food production in earlier times makes it more likely that famines occurred.   Changed institutional structures involving income and food entitlement would have affected the likelihood of famine in the past and in the future. Apart from institutional change, which is not always easy, greater and more regularized food supply reduces the incidence of famine.   Improved transportation and communication systems, international markets in food, international food aid, and emergency relief all contribute to further reduction of the incidence and impact of famine. The use of technology to create land and to provide a modest buffer against the vicissitudes of climate (such as irrigation to offset drought), also helps to reduce famine and/or malnutrition. Those types of changes have been taking place in the world in the last 3 to 4 decades. However tragic the losses to famine have been in recent years, overall lives lost to famine in the third quarter of the twentieth century are one-tenth of what they were in the last quarter of the nineteenth century, when the population was much smaller (Simon 1980, 1433). The frequency of famine in previous times is difficult for those of us living in affluence to imagine. Some have estimated that every year in China there was a famine in at least one region. In medieval Europe great strides were made in food production, even so, in Renaissance Britain famine still occurred about once every 10 years (Pirie 1976, 31).

The purpose here is not to defend modern technology by merely referencing past catastrophes that no longer occur or that occur infrequently. In discussing the quality of life, we are certainly concerned with more than just staying alive. Some sense of the qualitative dimension can be gained by studying technological changes that transformed lifestyles in the United States in the past century. Some argue that these technological changes have brought adverse changes in the quality of life and that some of the earlier technologies are prototypes of the kind needed for sustainable development. Further, the contrast between the earlier and contemporary technologies involves principles that are used to define the differences betwen soft and hard technologies or between mass and appropriate technology.

The three transformations discussed here are (1) technological changes in the household, particularly as a result of
the provision of energy and other services from centralized
sources; (2) the mechanization of transportation, particularly
the automobile; and (3) the increased life expectancy and the
impact upon our lives and institutions. Not all the gains in
these areas can be attributed to technology. However, these are
areas in which critics charge that technology and other aspects
of modernization have ruined our lives.

HOUSEHOLD TECHNOLOGY
    With calls for self-reliance and solar power, the soft
energy and appropriate technology movements are seeking to break
the household's dependence upon centralized energy sources and
manufacturing systems. If proper technologies are created there
may be merit in such future transitions. Though this is
doubtful, it is clear that the technologies and transformations
that created the modern household relieved its occupants of much
drudgery and adverse living circumstances.
    In Colonial America, according to Susan Strasser (1982,
32-61), houses "offered only scant protection from the winter's
ravages." In the late eighteenth century on into the nineteenth
century, cast-iron stoves produced in foundries improved home
heating. To keep out the cold, people tried to make their homes
as airtight as possible. Because heat and lighting sources
produced flames of various kinds, house fires were a constant
threat and the rooms were filled with smoke, soot, and poisonous
gasses. The soot and grime settled on the walls, the furniture,
and the people and necessitated laborious major cleaning every
spring. The air was uncomfortably dry and drafty. Central
heating, which greatly improved the situation, began to become
widespread at the turn of the century. It was dependent upon
cheap rolled steel and a national distribution facilitated by the
development of railroads.
    Cooking also first relied upon the hearth. "Before gas and
electricity, cooking was hard, hot, heavy and even hazardous."
writes Strasser. It was laborious, for it required constant
tending and was not as easily and exactly regulated as modern
stoves. Most homes did not have ovens until the twentieth
century (Bliven 1982, 104). Consequently, one of the virtues
ascribed to the nineteenth-century household, home baked bread,
was not as common as many believe. Rural people usually ate
pancakes instead of bread and the urban population bought their
bread (Strasser 1982, 34). Bread making at home appears to
be more of a common phenomenon today than in earlier times.
Citing the studies on "Middletown" in the mid-1920s and the
late 1970s, Bliven notes that 4 out of 5 housewives polled in
the recent study "regularly" baked their bread. In the earlier
study 2 out of 3 business class families and 4 out of 5 working
class families purchased their bread. Bliven also observes
that home canning did not become common until after 1900, when
factory-produced jars became available. "Until railroads,

refrigerator cars and canned goods made foods available year round, the American diet was monotonous--especially in winter-- and approached malnutrition. Not until the end of the nineteenth century did metallurgy improve so that pots and pans no longer rusted or released poisons into food," writes Bliven, concluding, "where preserving, cooking and baking in the home are concerned, the present is the good old days."

Along with stoves for heating, from the foundry came the cast-iron cookstove, which conserved energy and reduced the hazards of cooking with wood and coal. Wood, then coal, was the fuel source for heating and cooking. Candles were for lighting. During the nineteenth century a series of improvements produced coal-fired lockers for steam heat, oil and kerosene for lighting, and gas for heating and cooking. A later development, electricity, has been a boon to the household in uncountable ways in providing fractionalized power. Our dependence upon electric power becomes most obvious on those rare occasions that we lose it. Preeminent in the beginning, at least, was electricity for lighting. Previously, nighttime light sources were poor, not good for the eyes, and a hazard to the person and the home.

The transition in the United States from the nineteenth into the twentieth century was for the household to derive an increasing portion of the essential needs from industry or other centralized public or private sources. This was part of a transformation that few people would argue was anything but beneficial. There was also an increased use of the public water supply for drinking, cooking, cleaning, and waste disposal. The health benefits of this centralizing activity were enhanced by public health movements at the turn of the century that stressed, among other things, purification of drinking water. Waterborne diseases such as typhoid fever, cholera, and gastrointestinal ailments generally were among the leading causes of death in 1900 and are now practically nonexistent in the United States (Young 1961, 157; Galishoff 1980, 50-71). Rural water supplies also had problems of pollution in the nineteenth century (Bettmann 1974, 51). The provision of water in the household greatly lightened the burden of many tasks and freed the women from hauling hundreds of pounds of water (Strasser 1982, 103).

Naomi Bliven sums up the impact of these household changes, saying "that the industrial revolution emancipated many human beings from brute toil. Its greatest beneficiaries have been the poor. The spread of liberty and equality that under earlier technology had been no more than a philosophical phantom became a possibility." Feminism was one component in this pattern of movement to greater freedom, and technology has made possible for American families "a more egalitarian choice" of using machines to "manage a decent part of the housework." If Bliven's thesis is correct, these household machines tied to centralized technologies have been democratic technologies. Because they have greatly enhanced individual and family choice, they are also truly and operationally decentralizing. Those honest reformers who wish to break these technological nexus better be certain

that their small technologies can solve the problems as well as
the ones they seek to replace did.

FROM ANIMAL POWER TO HORSEPOWER
    Mechanized transportation possesses most of the charac-
teristics ascribed to alternate technologies. The railroads and
the horse-drawn and electric trolleys were transportation technologies for the masses who previously walked. In the late
nineteenth century the trans-Atlantic steamer was technology for
the masses of immigrants. In the twentieth century, the
automobile has become the decentralized transportation for the
masses (Brooks 1980, 71). In urban America, for millions of
people, the automobile was the first form of transportation they
owned other than possibly a bicycle.
    For urban areas, the horse-drawn carriage was inherently an
elitist means of transportation. It occupies a large space and
travels at best about 30 miles per hour. In addition, there are
problems in housing, feeding, and caring for large animals in
urban areas. In 1900, just as the automobile was coming into its
own in the United States, there were about 3 million horses in
urban America serving its population of just under 25 million
people (Census Office 1903, 40). This is less than 1 horse per 8
people. It includes horses used for all purposes: hauling,
stationary power, mounted police, and public conveyance. If
these numbers were subtracted from the total, we would see that
only a comparatively small proportion of the urban population had
access to personal horse transportation.
    Even with a limited population having access to horse-drawn
vehicles in 1900, the large number of horses in urban areas
constituted a service problem. Writes Tarr (1971, 65), "The
faithful friendly horse was charged with creating the very
problems today attributed to the automobile: air contaminants,
harmful to health, noxious odors, and noise." The horse was
rightly seen as a health problem that some thought the automobile
would cure. Each horse discharged gallons of urine and nearly
twenty pounds of fecal matter on the streets daily" (Melosi
1980, 15). The manure piles drew flies, gave off a foul odor
and formed a breeding ground for diseases. Streets were cess-
pools when it rained and pulverized dung blew into pieces in
dry weather. This manure dust was described as blowing "from
the pavement as a sharp powder to cover our clothes, ruin our
furniture and blow up in our nostrils" (Bettmann 1974, 3).
This powder irritated respiratory organs. Hauling away horse
carcasses was a major urban sanitation problem. Projecting
the exponential nineteenth-century growth of horse manure, many
humorous extrapolations found that "American cities would dis-
appear like Pompeii--but not under ashes," and by the mid-1970s
the entire United States would have been covered with several
yards of horse manure.
    One of the appeals of the automobile is its flexibility in
where and when we begin our journey and where we end it. It has
the kind of flexibility that Lovins finds appealing in soft-

energy technologies. Part of the flexibility of the automobile comes from the complementary, more centralized development of road networks and traffic control. Ironically, the critics of automobiles call for "mass transit," a more centralizing technology. This criticism implies that the automobile has become too decentralizing and too much a technology for the masses. Indeed, the problem of the automobile in the city is that there are so many of them that the resulting congestion limits flexibility and mobility. The social cost, in dedication of space and the destruction of neighborhoods, and the direct financial costs of building additional freeways and providing urban parking are considered prohibitive to many. The debate as to the merits and demerits of differing transportation systems can be left to the experts. The only point here is that though the analytical categories of soft, decentralized, etc., may be useful, they are certainly not definitive and do not close out inquiry or choice on issues of technologies. Nor are the social consequences of these choices of technologies rigidly fore-ordained in terms of the Lovins-type taxonomy.

TECHNOLOGY AND THE FAMILY

Many will concede the material comfort that science and technology have brought us but still question some of the intangible qualities of things that we have lost. The soaring divorce rate is cited as an example of the dissolution of institutions such as the family that is the result of the destructiveness of modern technology and modern life. Though the divorce rate seems to have stabilized (and might even be declining), few would argue that it is not a serious problem, particularly where the dissolved marriage involves children. But divorce is not the only means of dissolving a marriage; death serves the same function, only more tragically. In England a couple of centuries ago, in what Lawrence Stone (1977, 60) calls the Early Modern period, "after less than 17 years there was a 50% chance that the marriage would be broken by the death of [the man] or his wife." Most children suffered the loss of at least one parent before they became adults. Stone describes this situation as one "in which the family itself was a loose association of transients, constantly broken up by death of parents or children or the early departure of children from the home" (p. 81). Close to half the children would have died before their fifteenth birthday. It would be a rare family in which the child grew up without experiencing the death of a parent or sibling or that parents did not experience the loss of at least one child. Writes Stone (p. 5), "The expectation of life was so low that it was highly imprudent to become too emotionally dependent upon any other human being."

In the United States, according to Uhlenberg (1978, 78-79), "less than 10% of those born around 1870 survived to age 15, with both parents alive, and had all their brothers and sisters also survive to age 15." Infants had only a slightly better than 50% chance of reaching 15 with both parents alive. By 1950, over 90%

of the children born survived to age 15 with both parents alive
and over 80% reached 15 with both parents alive and all siblings
living until 15. These favorable trends continue to the present.
This change from 1870 to 1950 is attributed to an 88% decline
in the number of children who die during childhood and an 85%
decline in orphanhood among those who survive. Uhlenberg sums up
these statistics with the observation that "in less than 100
years the milieu shifted from one in which death within the
family was the usual childhood experience to one in which it was
the exception."

It is understandable that many students of the American
family deny that it is an institution in trouble or on the
decline. Some suggest that it is better than ever. Divorce
is a problem, but until 1960 the decrease in morbidity offset
any increase in the divorce rate. Bane (1976, 70) has argued
that "family disruption has not increased but has only changed
in character." People are married longer (first and second
marriages) and have more time together under circumstances
when interpersonal interaction is not complicated by other
commitments (i.e., children, work, etc.). Because of the
"decline in mortality, the orphanage has virtually disappeared
and grandparents have appeared" (Skolnick 1984, 32). Under the
circumstances of earlier centuries, it is not hard to understand
why some observers have found "no trace of affection in the
marital relationship." Stone repeats the French proverb, "If the
horse and wife fall sick at the same time, the . . . peasant
rushes to the blacksmith to care for the animal and leaves the
task of healing his wife to nature." The wife could be replaced
very cheaply, he says, while the family economy depended on the
health of the animal.

Recently in the United States we have been shocked to
learn how widespread sexual abuse and other forms of child
victimization are. We are horrified and rightly so, but however
bad the situation is, the community standards are such that we
are moved to investigate, learn as much as possible, and try to
correct the problem. In earlier centuries, the conditions for
children were worse and efforts to ameliorate them were sporadic
at best. One author (de Mause 1974, 1) has found that "the
history of childhood is a nightmare from which we have only
recently begun to awaken. The further back in history one goes,
the lower the level of child care, and the more likely children
are to be killed, abandoned, beaten, terrorized, and sexually
abused."

Of all the horrors delivered upon children, the ultimate is
death, infanticide. Previous chapters argue that advancing
science and technology give us moral options. In this context
it is noted that parents in impoverished circumstances had to
choose, consciously or unconsciously, which child to save and
which to let die. Sex ratios and other evidence indicate
that female infanticide (either directly or through relative
deprivation) is still practiced in some places. Terrible as it
may seem, infanticide was apparently widespread and common in

human history. Infanticide is seen by some as being a means for population control at times when other means were not available (Shorter 1973, 179). "In ancient times . . . infanticide . . . was a practice freely discussed and generally condoned by those in authority and ordinarily left to the decision of the father as the responsible head of the family," writes Langer (1974, 354). "Modern humanitarian sentiment makes it difficult to recapture the relatively detached attitudes of parents towards their offspring." In an earlier article Langer argued that even the seemingly benign use of a paid wet nurse was actually a disguised method of infanticide.

The varied means used to murder children down through the ages and in many cultures is so disgusting to the modern reader that it is almost beyond comprehension. In affluent industrial societies the practice has virtually disappeared and it is probably on the decline in most areas. Improved access to nutrition and effective contraception are removing major causes of infanticide. Infanticide is a scourge upon humanity that few would fault the progress of science and technology for their essential contribution to its elimination. Even the most strident proponents of the lifeboat ethic never argued for rejecting living members of one's own immediate family. If for most of humanity the lifeboat has been made large enough so that its newest most vulnerable members are no longer denied passage, it might be equally possible to build a lifeboat that will hold the entire family of man.

Privacy was also lacking in the early modern family, both among the poor and the rich (Stone 1977, 6). The prevalence of boarders made the nineteenth- and early twentieth-century American households also without very much privacy. Inadequate heating, which forced the occupants to huddle together in limited areas during winter, added to the lack of privacy. As Strasser put it, the "houses shrank during winter." Some may argue that the privacy that middle-class Americans grow up with creates loneliness. The isolation of the farm family, particularly in winter, created loneliness and boredom that was a concern to nineteenth-century American writers (Smalley 1893, 378-82).

Modern society is not without its problems. Some of our social problems could at least partially be the result of solving earlier problems. The marriage bond and the historically transient nuclear family become permanent as life expectancy increases. Stone suggests, not necessarily cynically either, that divorce is an "institutional escape-hatch" and "a functioned substitute for death." Presumably most people consider it a good thing that children grow to maturity and that parents are around to participate in or at least observe the process. Since long-term relationships were not all that common in earlier times (except possibly the late nineteenth and twentieth centuries), we may not yet have fully developed the means of having enduring, emotionally satisfying marriages. We should not ignore the fact that with all the reasons and opportunities for divorce, a large number of marriages seem to endure. Whatever the experts tell us

the problems of the modern family are, at least having people
alive gives us a basis and a hope that more stable relationships
can be established in the future.

Technology has benefited the poor. It is truly technology
for its masses. Mass technology functions best with mass
markets. Modern economists may differ as to whether the free
market or government fiscal or monetary policy are necessary to
maintain aggregate demand. All agree, however, that mass
technology means mass consumption. Affluent societies have
democratized the economy to a greater extent than other economies. Other technologies that imply limits to growth also imply
limits to participation in consumption. As was observed with the
horse and carriage, for urban areas there was a severe limit to
the portion of the population that could have this form of
transportation. Some of the limits-to-growth writers have been
quite candid about their elitism. The lifeboat theses and
triage are the most startling of the elite theories, but they
do follow logically if the overcrowded boat will drown us all or
if we all will starve. Triage and special privilege for elites
are supported by many who fear population growth. One author
(Clayton 1980, 175) states that poor countries are destined
to remain that way and that we have "to adjust to permanent
inequality." Population is described by another (Gregg 1955,
682) as a cancerous growth that demands food, but unfortunately
cancers "have never been cured by getting it." People are
described by a third (Stern 1955, 685) as an "epidemic." Ehrlich
(1971, 166-67) has also referred to human population growth as a
cancer and written of the many "brutal and heartless decisions"
this condition requires us to make.

There is a logical relationship between such ideas, however
shocking they may sound, and the limits-to-growth thesis. If
resources are fixed and finite and population is growing out of
control, then privilege, inequality, and some drastic controls
are necessary if anyone is to survive. It is not our intent to
tar all limits-to-growth, soft-technology, small-is-beautiful
supporters with the brush of elitism. However, since some of
them have claimed that their technology is uniquely democratic,
it is fair to point out that others operating from shared assumptions derive rather authoritarian, undemocratic conclusions. All
of this suggests that before we predefine technologies we ought
to examine their contemporary and historical consequences.

OVERCOMING LIMITS: TECHNOLOGY AND THE HUMAN POTENTIAL

If this narrative seems at times polemical, it is in
response to what are considered to be very highly polemical
attacks against modern technology. However, we do not wish to
battle an extreme perspective with an extreme view from the other
side. The vigor of this argument, or the sharpness of this
attack, should not blind the reader to the repeated recognition
of the real problems of the world and with modern technology. In
the problem-solving process described, the fact that modern
technology is better does not make it perfect. Not one thing

said in this volume justifies the use of any technology without regard to its long-term human and environmental consequences.

Modern technology has not failed us. It might be closer to the mark to say that we have failed it. This, of course, is meant figuratively, technology is our servant and not vice versa. Technology is a problem-solving process. As such, despite the many benefits it has conjured, we are far from achieving the full benefits for all humanity of technology as it now exists, let alone as it is continuing to develop. In this sense we have failed the potential of our technology. Many alternate-technology supporters consider their critique to be radical. If being radical is desired, then it is an equally radical criticism of political, economic, and social structures, no matter how progressive, if they leave unresolved problems that are readily solvable.

With this book an attempt has been made to create a conceptual framework for thinking about technology. A theory of technology is yet another human tool. In this case it is a tool for the very important process of selecting technologies for problem solving and selecting agendas for research. The principles laid out here are practical and useful to the actual development process, and it is hoped that they are read and interpreted in that light.

If we are to use technology sensibly, we have to be careful not to be rigid in the use of intellectual models. This book has tried to explore some of the implications of an evolutionary model. One motivation was the simplistic use of this model to support limits to growth. One author (Sale 1980) cited the great biologist J. B. S. Haldane to justify small as a governing principle. Haldane (1928, 18–26) does find biological advantages to smallness, but if the author had taken the trouble to finish this small essay he would also read where Haldane found advantages to bigness.

Some of the terms in the life sciences have taken on a public meaning well beyond their scientific use. Such is the case for ecology and coevolution. The term ecology defines a movement based on the interrelatedness of all living creatures, yet many within the movement favor small, supposedly self-reliant communities of human beings. Dependency, or interdependency, is treated as pathology or malignancy. The question is, if life forms are all interrelated, does this not apply to the human community? Coevolution is another term that has had social, intellectual baggage added to it. If symbiotic relationships are part of nature, and if these involve total dependence between different species, then does not this description apply to humans and the plants and animals that we refer to as domesticated? Is this more or less natural than the other coevolutionary systems?

However useful the evolutionary model, there are limita-tions. Many now see evolutionary fitness as being a matter of chance: an unforeseeable change in the environment can eliminate species that were previously well adapted (Beardsley 1983, 615). If current theories about periodic catastrophes are correct, then there is no adaptation that could prepare the life form for events that occur on the order of every 26 million years (Cherfas 1984, 28-30). Human societies, too, have sometimes found that well-adapted technologies become obsolete in changed circumstances of international competition. Some have failed to change and paid the price for doing so. The difference is that we have the choice of changing and adapting our technology.

These theories of periodic catastrophes are human theories; if they are correct it is because we have scientific inquiry and technological means to understand the past better and to gain increasing understanding of the shape of the future. We can preadapt and prepare for these events. We may not survive them but at least we have a chance. Of Earth's creatures, we are the only ones with a knowledge of life history, of the forces that have shaped us, and of the possibilities of the future. In this evolutionary activity of life, writes Day (1984, 255), "the most astonishing result of the whole process is that we know about it."

For the present, the discourse on changes in the environment must continue, but it must be embedded in scientific research and not in slogans. If our technology is a threat to survival, it is because it is effective in solving other problems. Those who would have us abandon and modify technologies need to frame the advocacy in terms of the entire extent of problems solved and problems created. Costs alone or benefits alone do not constitute a basis for intelligent choice. In responding to changing circumstances, the dynamics of technological change is an asset, not a liability. Technologies that others no longer use and on which no significant research is being undertaken are not likely to give poorer countries the capability to respond to change. Such technologies have been called "retarded technology" (O'Keefe 1983, 832) though the term is not meant as pejoratively as it sounds.

At various points in the process described, human beings and other life forms have faced limits. Had they lived within these limits as we have been counseled to do today, the process would not have gotten very far. We have survived by continuing to overcome limits. Living within limits too often has been the path to extinction. For the past few centuries the model of harmonies of the cosmos and of the ecosphere have had a great deal of intellectual appeal. They have served as a basis for much thinking in the social sciences. The alternate technology movements are very much predicated upon the harmonies of nature, particularly of the ecosystem. Yet scientific investigation has not entirely ratified this model. The ecosystem does not always have a set, harmonious equilibrium (May 1976a,b). Though competition is still a dynamic force in the ecosystem, "it may be no more important a determinant of the ecological structure than predation, parasitism, disease, and various random forces" (Simberloff 1984, 22). Evolution may not be as orderly as we have popularly interpreted it to be (Cherfas 1984, 28-30). Increasingly observers of the physical universe are studying chaos.   The physical universe is described as a rather violent place, with various forms of matter decaying.

Our system of ethics or values does not need nature as either a guide or basis for authentication. Experience has shown that if we get too much in the habit of rigidly reasoning from nature we will have to be concerned about any new trend in scientific inquiry. For instance, when the concept of the Earth as the center of the universe was essential to our self-image and value and religious systems, then post-Copernican astronomy was suppressed.   Our very real concern is that critics of modern technology and modern science are at times so shrill that we might dismiss them even when the problem that they are calling attention to is real.   Today there are important issues on which serious scientists are hesitant to speak because not-too-well informed catastrophists have already claimed the headlines (Boffey 1984, 19, 22; Budiansky 1984a, 301-2).

We have looked at how technology has helped humanity to overcome limits. We sometimes forget that this macro level is working because of microlevel changes. The poet told us about the "mute inglorious Miltons" lying in the village graveyard, never having the chance to develop this potential. As we expand opportunities, more human potential can be developed, from which all will benefit.   Samuel Florman (1981, 95) tells us about a live telecast in 1979 that featured soprano Leontyne Price and violinist Itzhak Perlman, with Zubin Mehta conducting the New York Philharmonic Orchestra.   The three were born in different parts of the world--Laurel, Mississippi; Tel Aviv; and Bombay-- and they studied and trained in many other places.   They were of different races and religious backgrounds and one, Perlman, cannot walk without crutches.   In earlier times, discrimination or other restraints would have kept them from practicing their art, let alone doing it together.   Florman asks, "What would have become of these individuals in a world that shunned 'bigness'?" Performances such as these are but one example of the vast reservoirs of human potential and the array of creative combinations possible in our world of modern science, technology, and communication.   It is this potential that must be continually expanded.   Whatever technological choices are made, we must make sure that we do not limit our vision as to what we are and what we may become.

BIBLIOGRAPHICAL ESSAY

CHAPTER 1

A Theory of Technology was in press when I encountered William Day's Genesis on Planet Earth (1984). Ironically, no single work upon which I based Chapter 1 comes as close to what I am trying to achieve as does Genesis on Planet Earth. I cannot recommend it too highly to anyone interested in the origins and development of life on earth and its implications for technology and society today. Brian J. Skinner's (ed.) Paleontology and Paleoenvironments (1981) is a diverse and useful book that includes some of the best writers on early life forms. The articles originally appeared in American Scientist and reflect the high quality and lucidity common to that publication. J. E. Lovelock's Gaia (1979) has become almost a cult book in some circles. His Gaia hypothesis that treats the totality of life on Earth and the atmosphere as a living, adapting being obviously appeals to those with cultist orientations. However, his thesis and supporting argument are useful in understanding the dynamics and interaction of the evolution of life and the environment. A scholarly work that finds the Gaia hypothesis to be "intriguing and charming but ultimately unsatisfactory" is Heinrich D. Holland's The Chemical Evolution of the Atmosphere and Oceans (1984). Another book read after the manuscript was complete was Not in Our Genes (1984), by R. C. Lewantin, Steven Rose, and Leon Kamin. Though their primary concern is with intelligence in human beings (and the controversial issues surrounding its origins and measurements), their closing statements on the way life forms create their own environment are both eloquent and replete with insight.

Two recent scholarly symposia, J. William Schopf (ed.) Earth's Earliest Biosphere (1983), and Yecheskel Wolman's (ed.) Origin of Life (1981), were vital for this chapter. Both are meant more for the specialist than for the casual reader. For an easy to read, profusely illustrated book on the subject, S. Dillon Ripley's (ed.) Fire of Life (1981), a collection of articles from Smithsonian magazine, is pleasing to the eye and the mind. Eric Chaisson's Cosmic Dawn (1981) is a masterful popular account that describes the origins of the universe and the origins and development of life on Earth, as well as present and future concerns. Cambridge Encyclopedia of Earth Sciences (1981), David G. Smith (ed.), is a reference work that has sections relevant to Chapter 1 and to Chapter 5.

The concept of entropy has fascinated thinkers for the past 150 years. Unfortunately, many overinterpret its significance and some seem to have a rather modest understanding of the issues involved. In my judgment, Jeremy Rifkin's Entropy (1980) makes grandiose claims about a "new world view" that are not warranted by the theory and facts presented. A far more thoughtful attempt to apply the concept of entropy to human affairs is Nicholas Georgescu-Roegen's The Entropy Law and the Economic Process (1971). Georgescu-Roegen tries to get around the fact that the sun is a source of negative entropy by treating the Earth and the human economy as energy and matter systems. I differ with his analysis but still find him intellectually provocative. P. W. Atkins in The Second Law (1984) presents a good history of the concept of entropy and its implications. Ilya Prigogine has raised questions about the inevitable long-term consequences for the physical universe implied by the theory of entropy. See Ilya Prigogine and Isabelle Stengers, Order Out of Chaos (1984). For the implications of Prigogine's work on entropy for social inquiry, see the articles by Wil Lepkowski, "The Social Thermodynamics of Ilya Prigogine," and "Science and the Humanities: Bridging the Gap," in Chemical and Engineering News, April 16, 1979, and December 1, 1980, respectively. For a good brief history of the theory of entropy see Peter Engel's review of the Prigogine and Stengers book, "Against the Currents of Chaos," The Sciences, September/October 1984.

The delights of the study of linguistics are nowhere more lucidly and simply presented than in Anthony Burgess, Language Made Plain (1965). The basic characteristics of language (as differentiated from signs or signals), such as reflexive and productive or open-ended, can be found there or in any other basic linguistics book.

Sherwood Washburn, "Tools and Human Evolution," Scientific American, September 1960, is a simple, popular, but nevertheless profound work on the interaction between human and technological evolution. It was an early but important influence on my views on technology. C. E. Ayres, The Theory of Economic Progress (1944), was an even earlier work that formed my thinking on technology. My later studies under Ayres and reading of his other writings, such as Towards a Reasonable Society (1962), formed a basic perspective that pervades this book.

James F. Feibleman is one of the few philosophers who has given serious and sustained critical thinking to technology and its human dimension. Too often what passes for humanistic or philosophical discourse on technology ends on a tangent that sets technology in opposition to the humanities and ethical concerns. Fortunately, in Feibleman we have a sophisticated and respected philosopher who has a broad comprehensive understanding of technology. Some of his writing on the subject includes

"Artifactualism," in Philosophy and Phenomenological Research, June 1965; "Importance of Technology," Nature, January 8, 1966; and Understanding Human Nature (1977).

An overall perspective on human evolution and prehistory can be gained by reading two fine textbooks. Both were recommended and lent to me by colleagues in anthropology. Though the term text is sometimes a pejorative and synonymous with boring, both are books that the intelligent, interested reader would find informative and pleasurable to read. They are Bernard G. Campbell, Humankind Emerging (1982), and Robert Wenke, Patterns in Prehistory (1980). C. Loring Brace, The Stages of Human Evolution (1967), is also a text, though more specialized, and is written by one of the distinguished students of human evolution. Walter Goldschmidt, Man's Way (1967), includes an analysis of technology from an anthropological perspective that complements Ayres' economist's view. Mark Nathan Cohen, The Food Crisis on Prehistory (1977), is a seminal monograph that is central to understanding the questions concerning the transition from gathering to domestication.

Man the Hunter (1968), edited by R. B. Lee and I. Devore, has long been considered a stimulating and essential work on the origins and development of humans as hunters. In recent years there has been some questioning into how early the characteristic hunting and gathering way of life began. Current estimates may reduce the length of time that humans have been hunters and gatherers below the 99% figure used in Chapter 1, but this way of life and its predecessor still constitute the overwhelming percentage of time of human existence. See Richard Potts, "Home Bases and Early Hominids," American Scientist (July-August 1984), and for a popular survey of the new interpretation and evidence see Erik Eckholm, "Theory on Man's Origins Challenged," New York Times (September 4, 1984).

As a graduate student, I received as a handout a copy of a paper (article or lecture?) by Jack Harlan, "Crops, Weeds, and History." Over the years, this small work has strongly influenced my ideas on domestication of plants and, more important, on the dynamics of change generally. Only in research for this book did I encounter a much larger work by Harlan, Crops and Man (1975), that incorporates many of his provocative ideas. A more recent and equally stimulating monograph with excellent sections on domestication is by J. G. Hawkes, The Diversity of Crop Plants (1983). Ester Boserup, The Conditions of Agricultural Growth (1965), has not had the profound impact upon economists that it has had among anthropologists, particularly those working on historical and/or developmental problems. Her work on population growth and changes in agricultural technology is original and gives rise to fruitful inquiry whether or not one agrees with her thesis.

CHAPTERS 2 AND 3

These two chapters bear a heavy debt to joint research and publication carried out by me with Oriol Pi-Sunyer. Of the communications, reviews, and book that we co-authored, material (including several successive paragraphs) was taken verbatim from "Cultural Resistance to Technological Change," Technology and Culture, Spring 1964, and "Technology, Traditionalism, and Military Establishments," Technology and Culture, Summer 1966. Most of the ideas about technology expressed in this book evolved out of the author's reading and experience in a manner similar to our description of technological evolution. Chapter 2 represents my earliest thinking about technology and its implications, and consequently the sources used reflect research current at that time. In its time, Abbot Payson Usher's A History of Mechanical Inventions (1959) was a monumental empirical study of a vital stream of technology in Western culture. Though much of the research may be somewhat dated, the book contains numerous insights on the nature of invention and technological processes. Another founding father of the modern study of the history of technology is Lynn White, Jr. His Medieval Technology and Social Change (1962) is a creative effort in the tradition of the great French historians, such as Marc Bloch, showing the way in which technological change can be the causal element in a much larger social, political, and economic transformation. M. C. Duffy's "Mechanics, Thermodynamics, and Locomotive Design" (1983) reflects the way in which the history of a teehnology can also be understood as a history of ideas.

Thorstein Veblen, in his voluminous writings, such as The Instinct of Workmanship (1922), found technology to be the dynamic force in economic development. Veblen's influence on C. E. Ayres and others laid the foundation for Institutional Economics, a school of economic thought in which the study of technology is a necessary part of understanding economic processes. Institutional economists generally are affiliated with the Association for Evolutionary Economics, which publishes the Journal of Economic Issues. Jacob Bronowski, Science and Human Values (1956), is a work that inspires awe. Here is a great scientist who looked at painting and literature and demonstrated more than anyone else in our time the compatibility of science, humanistic endeavors, and democratic traditions. His was a breadth of vision and a range of interdisciplinary inquiry that was set forth in this small, unassuming, but very fine book.

Over the past two decades, few if any economists have written as extensively and intelligently on technology as has Nathan Rosenberg. Much of his writing has been scattered about in various scholarly journals. Fortunately for the not always diligent researcher, his pieces are periodically collected and issued in book form and almost always appear as fresh as when they were first published. Inside the Black Box (1982) is the latest of several books.

Chapters 2 and 3 raise the issue of the direction of causality between science and technology. Rosenberg was one important source, as was A. R. Hall, The Scientific Revolution (1962; later editions are available). On the central importance of technology for science, the work of Edwin Layton, "Technology as Knowledge," Technology and Culture, January 1974 (one of several articles of his on the subject), and Derek J. de Solla Price, "Of Sealing Wax and String," Natural History, January 1984, are the most important on the subject. The influence of technology and mechanical processes on scientific theory can be seen in John Cornell's "Analogy and Technology in Darwin's Vision of Nature," Journal of the History of Biology (1984). Any attempt to understand this issue in technology would require the reader to survey the roughly 25 years of the journal Technology and Culture. A new journal, History and Technology, will apparently address these questions, as is evident from an article in vol. 1, no. 1, Jan Sebestik, "The Rise of Technological Science," 1983. An excellent survey article on technology and science, in which such eminent historians of technology as Melvin Kranzberg are interviewed, is William J. Broad, "Does Genius or Technology Rule Science?" New York Times, August 7, 1984. Every Tuesday the New York Times has a section called Science Times. It could more accurately be called Science and Technology Times, with articles that are useful indicators of important work and discoveries.

CHAPTER 4
Chapter 4 was rewritten from earlier pieces for inclusion in this book. Most of the material originally appeared as two different full-page Op-Ed pieces in the Houston Chronicle. The entire chapter, except for some later changes and additions, was included in The Apocalyptic Vision in America (1982), Lois Parkinson Zamora (ed.).
Paul Ehrlich, more than any other author, is most closely identified with apocalyptic visions of impending catastrophe. Ehrlich's The Population Bomb (1971) was widely read and cited. His articles appeared in a range of publications, many of which are cited in this book. Particularly significant are the lurid articles "Eco-Castrophe," Ramparts (September 1969); "World Population," Reader's Digest (February 1969); and "Countdown to Disaster," Wall Street Journal (December 3, 1968). Less well known but equally catastrophic in vision is William Paddock and Paul Paddock, Famine 1975 (1967). In an act of extreme self-assurance that was totally unwarranted by any verified prediction, the Paddocks allowed their book to be reprinted without change in text as Time of Famines (1976). That their forecasts had fortunately failed to materialize merely led the Paddocks to proclaim them more stridently and to chide their detractors in a preface to the new edition. Of all the widely read doomsayers, Garrett Hardin in The Limits of Altruism (1977) most closely attempts to use scientific methods. In fact, his essay in this volume, "Vulnerability--the Strength of Science," would be accepted by most theorists as a lucid, fair statement of scientific methods and philosophy. A number of his writings on population control and the limits of "the commons" have a wider audience beyond the catastrophists. Hardin is also about the most overtly elitist of the apocalyptic writers. Stephen

Schneider, a meteorologist, is more restrained but still highly pessimistic in his The Genesis Strategy (1976). His recent book with Rondi Londer, The Coevolution of Climate and Life (1984), is scholarly and looks at past, present, and likely future climates. Schneider and Londer demonstrate the complexity of long-term climatic forecasting models, and it would be fair to say that they are more than a little concerned with the changes introduced by the modern technological economy. The earlier sections on the "coevolution" of climate and life are compatible with the authors we used in Chapter 1.

Donella H. Meadows et al., The Limits to Growth (1972), was generally referred to as the "Club of Rome" study and was widely cited for its "proof" that we were rapidly exhausting the world's mineral resources. The Club of Rome was a group of primarily European businessmen and intellectuals concerned with large issues of humanity, society, and the future. The Limits to Growth study was sponsored by the Club of Rome and used the "systems dynamic" approach of Jay Forrester. A little over a decade after publication, few economists or others give any credence to the long-term forecasts of The Limits to Growth.

The Club of Rome sponsored another study on the Earth's limits, Computation of the Absolute Maximum Food Production of the World (1975), by P. Buringh, H. D. J. van Heemst and G. J. Staring, which clearly shows that we are not anywhere near the Earth's limits in food production. For the purpose of their study, they introduce a variety of unrealistic assumptions, all clearly stated, such as utilization of every day of the growing season on all arable lands, but the magnitudes of the potential increases in food supply are so great that even substantial modification of the assumptions would not alter the conclusions. Further, they used the assumptions of constant technology that biased the outcome in the opposite direction.

Two early and excellent critiques of the doomsday analysis are John R. Maddox, The Doomsday Syndrome (1972), and H. S. D. Cole et al., Thinking About the Future (1974). The latter was published in the United States as Models of Doom (also 1974).

Though it was the idea of mineral exhaustion that generated the first round of writing on limits in the early 1970s, it was the "energy crisis" of the mid and late 1970s that gave the limits to growth and small-is-beautiful literature a wider audience. Two of the more sensible works on energy, its potential development, and more efficient utilization (including conservation) are Robert Stobaugh and Daniel Yergin (eds.), Energy Future (1979), and James Griffin and Henry Steele, Energy Economics and Policy (1980).

The best comprehensive research and writing on the CO_2 problem in the atmosphere is by the National Research Council of the National Academy of Sciences (U.S.). The latest large report is Changing Climate (1983). Although the primary purpose of a book review is to tell about the book, occasionally one or two are so good that they relate more about the subject than most articles. Such is the case on CO_2 for Sherwood R. Idso, "Through a Glass Darkly," New Scientists (April 19, 1984), and John S. Perry, "Much Ado about CO_2," Nature (October 18, 1984). The books they reviewed are included in the reference list. There are varied interpretations as to the impact of deforestation upon CO_2 in the atmosphere. Two examples are "Global Deforestation," Science (December 9, 1983), by G. M. Woodell et al., and "Global Forests," by Roger Sedgo and Marion Clawson, in Julian Simon and Herman Kahn (eds.), The Resourceful Earth (1984). The process of deforestation is seen as a causal factor in other atmospheric and climatic change in Bayard Webster, "Forest's Role in Weather Documented in the Amazon," New York Times (July 5, 1983).

CO_2 is not the only change contributing to atmospheric warming from the greenhouse effect. There is continuing concern from chlorofluorocarbons and now from methane gas. Some of the latest findings are reported in Richard A. Kerr, "Doubling of Atmospheric Methane Supported," Science (November 29, 1984). Methane emissions, however, may help to protect the ozone layer where there is renewed concern for its stability, as reported in H. I. Schiff, "Ozone Fears Revisited," Nature (November 15, 1984) and Michael J. Prather, Michael B. McElroy and Steven C. Wofsy, "Reductions in Ozone at High Concentrations of Stratospheric Halogens," Nature (November 15, 1984). The CO_2 controversy has led to an interest in the entire carbon cycle, the search for "missing" carbon, and the extent to which water (rivers, lakes, oceans) are a "sink" for the cumulation of carbon. A good survey of the current state of this research is Egan T. Deggens, S. Kempe, and V. Ittekot in "Monitoring Carbon in World Rivers," Environment (November 1984).

CHAPTER 5

The fascinating story of the creation of the elements larger than hydrogen and helium in the furnaces of stars can be found in many sources. Three very good, readable books including this topic are David H. Clark, Superstars (1984), George A. Seielstad, Cosmic Ecology (1983), and Hubert Reeves, Atoms of Silence (1984). There would be no minerals and no life as we know it without the birth and violent death of stars. A more detailed and technical account of the creation of elements is William A. Fowler, "The Quest for the Origin of the Elements," Science (November 23, 1984). Fowler shows the complexity of the problem but still finds "that each one of us and all of us are truly and literally a little bit of stardust."

The year 1984 saw the publication of two first-rate books with essentially diametrically opposed perspectives on resource exhaustion, environmental degradation, and a host of other related issues (a number of which we cover in this book) which gives us a sense of the direction in which the global economy is headed. The Resourceful Earth (1984), by Julian Simon and Herman Kahn (eds.), is, as its subtitle suggests, a response to a "Global 2000" report. The many authors of this volume rely heavily on trend data to find the prospects for the human condition to be improving. Simon's introduction (read by Kahn before his death) is very assertively optimistic, while the authors of the essays are more cautious but still optimistic. Lester Brown is a prolific author who has been warning us of long-term dangers in our current technological practices, particularly in agriculture. State of the World, 1984 (for which Brown is senior ed.) finds some areas of improvement, but many areas for environmental concern. Worldwatch, the publication's sponsor, is an environmental group that has a strong interest in third world development and is generally favorable to small technologies. State of the World, 1984 is presumably the first of a series of annual reports.

The Global 2000 Report to the President (1980) is a multi-volume study edited by Gerald O. Barney; it offers a pessimistic assessment of resource availability in the year 2000. Though we find, along with Simon et al., its assessment fundamentally in error, many of the creative ideas for conservation and resource recycling are useful and possibly economically justifiable apart from any expectation of resource exhaustion. The Global 2000 Report bears some similarity to the massive Paley Commission report, Resources for Freedom (1952), though the Paley Commission was less global in its concerns and less ideological in its orientation.

The book that more than any other has set the agenda for the past two decades for the economist-discussion of mineral resources is Harold J. Barnett and Chandler Morse, Scarcity and Growth (1963). Even more basic to this chapter is the work of Erich W. Zimmermann, World Resources and Industries (1951). This massive work of theory and empirical research is largely out of date in its data but still fresh and original in its opening theoretical chapters. Nathan Rosenberg, one of the best economists regularly writing on technology, has useful ideas on technological progress and resource availability in works such as "Innovative Responses to Materials Shortage," American Economic Review (May 1973). Charles Maurice and Charles W. Smithson in The Doomsday Myth (1984) argue that throughout 10,000 years of human history, resource crises have been resolved through markets if the price system is has been allowed to function. One of the best collections on science, technology, and resources from the perspective of the research scientists and technologists is Materials: Renewable and Non-Renewable Resources (1976), a compilation of articles from Science, by Philip Abelson and Allen L. Hammond (eds.).

The possibility has been raised of deriving minerals and energy from extra-terrestrial sources. The best study on minerals from the moon is David R. Cresswell, Extraterrestrial Materials Processing and Construction (1980). The mineral resources of the solar system are studied in three fine books, Elbert King, Space Geology: An Introduction (1976); Billy P. Glass, Introduction to Planetary Geology (1982); and John S. Lewis and Ronald G. Prinn, Planets and Their Atmospheres (1984). The potential of orbiting satellites for processing these minerals, for manufacturing, and for concentrating and beaming to

Earth solar energy has had a leading advocate in Gerard O'Neil. See his "The Colonization of Space," Physics Today (September 1974), and High Frontier (1977). See also a scholarly technical journal, Space Solar Power Review (vol. 1, nos. 1 and 2, were issued in 1980).

In the text we cited David Morawetz, Twenty-five Years of Economic Development, 1950–1975 (1977), in indicating that the direction of change in the world's economy has not been towards doomsday, but away from it.

CHAPTER 6

The journal Technology and Culture has published numerous articles on the influence of technologies on cultural development. No research on technology that has any historical dimension is complete without a thorough survey of Technology and Culture. Two articles from this journal of particular importance to this chapter were Barrington Nevitt, "Pipeline or Grapevine" (April 1980), and Arthur D. Kahn, "Every Art Possessed by Man Came from Prometheus" (April 1970). No technology is more often cited for its impact upon Western civilization than printing by movable type. Of all the many fine volumes that I have read on printing's origin and impact, none compared in the quality of scholarship, in imaginative insights, and in intellectual excitement to Elizabeth L. Eisenstein's monumental tome, The Printing Press as an Agent of Change (1979) or the later, shorter version, The Printing Revolution in Early Modern Europe (1983). For concise, intelligent insight into the development of printing by movable type, the relevant chapters in Abbot Payson Usher's History of Mechanical Inventions (1959) are still first-rate. The impact of the technology on book publication and intellectual inquiry can be found in Eugene S. Ferguson's thoughtful and provocative article "The Mind's Eye," Science (August 26, 1977). For the large historical understanding of the relationship between technology and the arts, few have achieved such consistently high standards as Cyril Stanley Smith in such works as A Search for Structure (1982). David P. Billington has been exploring and clarifying the positive relationship between engineering and the arts in books and articles such as The Tower and the Bridge (1983) and David Billington and Robert Mark, "The Cathedral and the Bridge," Technology and Culture (January 1984).

Anthropologists have long been concerned about technological diffusion and cultural change. The literature is plentiful and replete with controversies, making it difficult for a non-anthropologist to cite a representative sample. Several of the works most appropriate to this chapter were encountered by me after this manuscript went to the publisher. David E. Whisnant's All That Is Native and Fine (1983) is an intellectually stimulating work on the potential pitfalls of outsiders imposing their interpretation of a culture in the name of preserving it. That this book deals with the Appalachian region of the United States in the first half of this century makes it no less relevant to questions about preserving cultures in the third

world today. In fact, I know of no work on the third world on
this subject that is more important than Whisnant's. Another
fine book, The Invention of Tradition (1983), edited by Eric
Hobsbawm and Terence Ranger, has essays on different countries
and cultures, showing that much of what is thought of as ancient
tradition is actually of comparatively recent vintage and all
too often an import. An interesting and balanced study on what
is called the anthropology of tourism is Valence L. Smith,
Hosts and Guests (1977). William Rubin's "Primitivism" in
Twentieth Century Art (1984) is a two-volume book accompanying
an art exhibition that demonstrates clearly the diffusion and
positive impact of art from the third world to the industrialized
countries. Five excellent articles on the exhibition and its
significance are Thomas O'Neil, "The Primitive Urge," Horizon
(September 1984); Douglas C. McGill, "What Does Modern Art Owe to
the Primitive?", New York Times (September 23, 1984); John
Russell, "Primitive Spirits Invade the Modern," New York Times
(September 28, 1984); Paul Richard, "Magical Affinities,"
Washington Post (September 30, 1984); and Manuela Hoeltexhoff,
"The Call of the Wild," Wall Street Journal (October 9, 1984).

In many instances small ethnic groups are being victimized
by larger, more technologically advanced societies. Where lands
are being forcibly taken, where societies are being wrenched
apart by external imposition, and where a people are on the verge
of annihilation, their preservation takes on a different meaning.
We should always distinguish between this type of circumstance
and the externally-imposed preservation activities that seek
to prevent a people from voluntarily undergoing change. Two
examples of scholarly work that intellectually complements
practical efforts to protect the rights of small groups are The
Geological Imperative (1972), by Shelton H. Davis and Robert
Mathews, and Shelton H. Davis, Victims of the Miracle (1977). A
more recent work that surveys small scale societies under stress
from rapid change is Jean E. Jackson, "The Impact of the State on
Small Scale Societies," Studies in Comparative Development,
Summer 1984.

Many critics of modern technology and science argue that
science and technology stifle the imagination. This chapter
has tried to show how advances in science and technology have
facilitated artistic expression. The argument is made that
modern technology has enabled scientific inquiry that is in and
of itself an act of creative imagination equal to that of any of
the arts. This statement of appreciation of science as art is
made by one who greatly loves what is more traditionally called
art and literature.

Two contemporary American poets, Ariel Dorfman, "Bread and
Burnt Rice," Grassroots Development (1984), and Frederick Turner,
"Escape from Modernism," Harpers (November 1984), find positive
relationships between technology and the arts and cultures.
Dorfman sees development and new technologies as creating
opportunities for cultural expression in Latin America. Turner
sees modern science and technology as forging new unity in our

cultural understanding and new technical means for artistic expression. Turner's subtitle, "Technology and the Future of the Imagination," is an affirmation that the powers of imagination are being enhanced. We illustrate science as imagination with a brief passage on cosmology and the origins of the universe. Those of us who read the popularizations on this subject have to read it as literature, for the basic inquiry is so technical and complex that even the most lucid popularizations can give us only an illusion of understanding.

Steven Weinberg's The First Three Minutes (1977) has in many ways set the standard for other books on the origin of the cosmos. A more recent book, James S. Trefil's The Moment of Creation (1983), discusses the theories on the origins of the universe and carries the inquiry up to the present. The Very Early Universe (1983), edited by G. W. Gibbons, S. W. Hawking, and S. T. C. Sikos, reflects the complexity of the inquiry and has a few scattered passages accessible to the lay reader. Not only can most of us not comprehend these incredibly small time periods (i.e., 10 to some minus power) or very long periods (i.e., 10 to some positive power), we are baffled as to how the inquiry has led to these theoretical observations. On the subject of extremely small to very large distances, the layman's ability to comprehend is greatly enhanced by a delightful popularized, illustrated book, The Powers of Ten (1982), by Philip Morrison, Phyllis Morrison, and the Office of Charles and Ray Eames. However limited our fundamental comprehension may be, to read these works in cosmology and astrophysics does expand our imagination and insight the way other artistic triumphs do. It makes us more aware and open and imaginative in the particular intellectual endeavors in which we are engaged.

CHAPTERS 7 AND 8

The opening pages of this section originated in a communication to Technology and Culture in response to a piece by John Bryant, "Systems Theory, Survival and the Back-to-Nature Movement," Technology and Culture (April 1980). On the comparative risks from natural hazards, two works are preeminent, Robert Kates, Risk Assessment of Environmental Hazard (1978), and Judith Dworkin, "Global Trends in Natural Disasters, 1947-1973," a working paper (1974). Kates and other authors refer to "technological hazards." In this regard see Robert C. Harris, Christopher Hohenemser, and Robert Kates, "The Burden of Technological Hazard," in G. T. Goodman and W. D. Rowe (eds.), Energy Risk Management (1978). Recent critics argue that current patterns of development increase the risks from natural hazards, as do many of our attempts at post-disaster relief. The most detailed exposition of this thesis is Frederick C. Cuny, Disasters and Development (1983). A more popular presentation of this idea, drawing from Cuny, is Erik Eckholm, "Fatal Natural Disasters on Rise," New York Times (July 31, 1984). Some specific examples of these new risk conditions can be found in Debora McKenzie, "Man-made Disaster in the Philippines," New

Scientist (September 13, 1984). Others argue that modern
agricultural technology and regional diversification have reduced
the risk of food loss; see Robert S. Loomis, "Agricultural
Systems," Scientific American (September 1976).

A recent forceful statement of the thesis that current
patterns of development are significantly increasing the risks
from natural disasters is receiving considerable media attention
as this book goes into production.   There was a wire service
story (Joan Mower, Associated Press, November 14, 1984),
editorial references to it, and articles in the Washington Post
(November 14, 1984) and the New York Times (November 18, 1984).
The work is Natural Disasters (1984) by Anders Wijkman and Lloyd
Timberlake.   I obtained a copy and found it to be one of the
best statements on the position critical of the use of modern
technology and the patterns of economic development.   Few would
argue with the study that there has been carelessness in many
areas in the use of technology or that development and population
growth have led to serious problems of deforestation, erosion, or
other forms of environmental degradation in many parts of the
world.   However, granting as we have throughout the book that
there are problems to be solved (and always will be) does not
validate their theses that overall conditions of life are
becoming more hazardous or that modern technology is not the
potential source of solutions to these problems.   Natural
Disasters is published by Earthscan, which also publishes a
variety of other works on environmental problems and on
technologies (generally small) for development problems.

Another story that has gained media attention is the
Ethiopian famine.   The situation is truly tragic and the need
for assistance is great.   The drought that brought on these
terrible conditions is widespread throughout Africa, as is
food deprivation.   However, in one of the countries affected,
Zimbabwe, "specialized seeds" and other agricultural imputs
allowed indigenous peasant farmers to make use of the small
rainfall that did occur to produce a record crop and for the
large commercial farms to produce "remarkably high yields."
Zimbabwe's achievements are noted by Glen Frankel in "An African
Success Story" (Washington Post, November 20, 1984) in a series
of lengthy articles on "Africa: The Hungry Continent."   Henry
Kamm, "Zimbabwe Beats Drought with Buoyancy and Skill" (New York
Times, December 2, 1984) also notes the record peasant harvest in
Zimbabwe despite the drought.   Illustrative of disparities in
food production within Africa and between Africa and the rest of
the world was another article on the same page by Seth S. King,
"Worldwide Food Production Levels Show a Big Rise" (New York
Times, December 2, 1984), and a large picture of the situation in
Ethiopia.

As noted in the text, trends in food supply in Africa have
been moving counter to the favorable worldwide trends.   There is
justifiable fear that when the publicity fades, assistance will
falter.   Even if relief is sustained until rains and a good
harvest return, there should be even greater fear that both

donors and local governments will not take the necessary actions
in science and technology in agriculture to bring greater food
security and significantly reduce the possibility that such a
tragedy can occur again.

If one looks at the figures for lives lost, as stated in
Natural Disasters, or as being estimated for Ethiopia, it is
clear that they are quite small compared to those from famine and
other disasters earlier in this century or in prior centuries.
Obviously, both historic and contemporary calculations are
subject to substantial errors of definition and estimation, but
the magnitude of the difference is sufficient to warrant an
assertion that contemporary losses are lower. These statements
are made not to minimize the urgency of action today but to
reinforce those whose compassion has been aroused, with the
hope that help is not futile and that the problems of hunger
are soluble. Worldwide average life expectancy has reached
unimagined length (about 60 years), so it is difficult to make
the case that life today is more hazardous. Again, this is not
meant to argue for complacency but to motivate action to raise
the average by lowering the death rate of those whose life
expectancies are considerably below current averages. A small
work just published sets an agenda for action on world agriculture, resources, the environment, and the quality of life.
The Global Possible (1984) is a study that resulted from a May
1984 conference sponsored by the World Resources Institute and
included distinguished representatives from many disciplines and
a broad spectrum of perspectives.

One of the risks associated with modern economies concerns
the lack of diversity in agricultural ecosystems. Some work in
modern ecology challenges the presumed greater stability of more
complex systems. An outstanding work on this subject is Robert
M. May, Stability and Complexity in Modern Ecosystems (1973).
There is a related view questioning modern agriculture from the
perspective that "nature knows best." Rene Dubos, The Wooing of
the Earth (1980), presents a delightful defense of the human
transformation of the environment and gives instances where
nature has failed. Some examples of nature's "failures" both for
prior times and for the present can be found in Dubos and in
"Algae: Mass Extinction," New York Times (May 10, 1983), and
Philip Hilts, "Bird Population of 17 Million Vanishes," in both
The Guardian (London) and the International Herald Tribune
(Paris) on March 16, 1983. Dubos argues that at times humans
have increased biological diversity and completed work (such as
recycling) that nature left unfinished. In addition, Julian
Simon and Aaron Wildavsky, "On Species Loss, the Absence of Data
and Risks to Humanity," in Julian Simon and Herman Kahn (eds.),
The Resourceful Earth (1984), argue that seed breeding and
storage techniques allow for the addition of new species without
the loss of old ones, which has been the historic experience.
Similarly, modern breeding methods are helping to sustain species
in zoos. On this point, see Jeremy Cherfas, "Test-tube Babies in
Zoos," New Scientist (December 6, 1984).

Julian L. Simon in "Resources, Population, Environment," Science (June 27, 1980), argues forcefully with statistical support that the hazards to human life are far less in recent times than they were in the past century or before. These arguments and data are repeated in Simon's The Ultimate Resource (1981), where he argues that the most important of Earth's resources is the intellectual creative power of human beings. Simon also expresses the idea that humans create arable land, as does Theodore W. Schultz, Investing in People (1981).

John Lenihan and William W. Fletcher have edited a series of useful volumes on a range of issues involving man and the environment. Particularly useful for this chapter is their volume, Food, Agriculture, and the Environment (1976). Another extremely important work is David Pimentel and Marcia Pimentel, Food, Energy and Society (1979).

Consumer Reports has regularly sought to undermine the mythology of so-called natural foods and the substantially greater price paid for food labeled natural or organic. One of the best overall articles on the subject is Martha E. Rhodes, "The 'Natural' Food Myth," The Sciences (May/June 1979). Two short popular articles critical of natural food fads are Jane E. Brody, "How Wholesome is 'Natural' Food," New York Times (December 12, 1979), and Elizabeth Whelan, "Beyond the Labels," New York Times (January 14, 1980). Harold McGee, On Food and Cooking (1984), explores the "science and lore of the kitchen" and seeks to demythologize our conception of foodstuffs and their preparation. A similar inquiry can be found in Joseph Hulse, "Food Science and Nutrition," Science (June 18, 1982). For a history of the plants (including flowers) that transformed the environment of the United States and formed the basis of agriculture and foodstuffs, see the charming and fascinating account by Claire Shaver Haughton, Green Immigrants (1978).

A. Carl Leopold and Robert Ardrey, "Toxic Substances in Plants and Food Habits of Early Man," Science (May 5, 1972), give an account of the substances in the food that our ancestors ate, including many foodstuffs that are still part of our diet today. Since plants are not mobile, toxicity is one of their few means of defense against predators. Most of the articles in the previous paragraph indicate some of the toxics in our contemporary diet. Edith Efron, The Apocalyptics (1984), sees the foods we consume, and not the man-made chemicals in them, as a significant cause of cancer today. Her book is massive and documented, but her tone is shrill and her argument is marred by implications of dishonesty ("the Big Lie") by those with whom she disagrees. Louis Lasagna, "Masked Men of Science," The Sciences (Jan/Feb 1984), is supportive of Efron's theses on the causes of cancer and her critique against the cancer research scientists. See also my review of Efron's book, "Cancer and Testing," Houston Chronicle (August 5, 1984). A differing perspective from Efron's can be found in Erik Millstone, "Food Additives: A Technology Out of Control?" New Scientist (October 18, 1984).

Bruce Ames, "Dietary Carcinogens and Anticarcinogens,"

Science (September 23, 1983), has done probably the most systematic and massive work on the carcinogenic and anticarcinogenic properties of foodstuffs. My use of the Ames article is not intended as an excursion into the argument among scientists on the proportions of human cancer caused by "natural" food and by man-made chemicals. A balanced assessment of a particular likely carcinogen, formaldehyde, that is both natural and at times man-made can be found in Fred Pearce, "Coming to Terms with a Carcinogen," New Scientist (October 18, 1984). A nontechnical presentation of Ames' work can be found in Janet Hopson and Joel Gurin, "Diet and Cancer--Round 2," American Health (November 1984). Few writers are as much fun to read as the late Marston Bates, and an interest in eating is a good excuse to read another book by him, in this case Gluttons and Libertines (1967).

Those interested in the way that technology creates land on the frontier or even defines the frontier of a society may read James Scobie, Revolution on the Pampas (1964), or my own work, Technology and the Economic Development of the Tropical African Frontier (1969). Another source is Walter Prescott Webb, The Great Plains (1931), though Webb in other works on his frontier thesis explicitly denies that technology can create land or resources. On the potential of science to open up the Amazonian frontier, three works stand out among many fine ones: Pedro Sanchez et al., "Amazon Basin Soils," Science (May 21, 1982); Emilio F. Moran (ed.), The Dilemma of Amazonian Development (1983); and S. B. Hecht (ed.), Amazonia (1982). Moran has published extensively on Amazonian development. A more cautious, if not pessimistic, assessment of the impact of deforestation in the Amazon on the water cycle can be found in Eneas Salati and Peter B. Vose, "Amazon Basin," Science (July 13, 1984), and on species loss can be found in Roger Lewin, "No Dinosaurs This Time," Science (September 16, 1983).

Of the writers who have expressed concern and in some respects alarm at the continuing shrinkage of tropical rain forests, none is more coherent and sensible than Norman Myers. In his latest book, The Primary Source (1984), he argues for "full sustainable use of the tropical forests" without having to cut them down. Similarly, in The Sinking Ark (1979) Myers makes a case for the "utilitarian benefits" of preserving and protecting plant and animal species from extinction. In these and other works Myers recognizes that there are economic pressures and necessities in countries where vulnerable forests and species are, and that some strategy short of 100% preservation will have to be adopted if competing needs are to be reconciled. Two authors who have been active and written voluminously on world environmental and food issues, generally in alarmist tones, are Erik Eckholm and Lester Brown. Representative of their extensive work is Eckholm's Losing Ground (1976) and Brown's (written with Eckholm) By Bread Alone (1974). The issue that Lester Brown, Erik Eckholm, and a number of others raise concerning the loss of soil from farmland and the loss of farmland to development is questioned for the United

States in Pierre R.   Crosson (ed.) <u>The Cropland Crisis</u> (1982);
John Baden (ed.) <u>The Vanishing Farmland</u> (1984); and Pierre
Crosson, "Agricultural Land: Will There be Enough," <u>Environment</u>
(September 1984).

The critical question of soil erosion is addressed in "The
Threat of Soil Erosion to Long-term Crop Production," <u>Science</u>
(February 4, 1983), by W. E. Larson, F. J. Pierce and R. H.
Dowdy.   Two contrasting views on soil erosion can be found in
Theodore Schultz, "The Dynamics of Soil Erosion in the United
States," in the Baden (1984) book, and in Lester Brown and Edward
Wolf, <u>Soil Erosion</u> (1984).

<u>Important</u> in understanding issues and controversies over
agriculture and the changing of the atmosphere or the terrestrial
ecosphere are the cycles of the various elements, such as carbon,
nitrogen, and sulfur, and of water. The way that these and other
constituents of life work their way through this ecosystem—the
air, the rivers and oceans, and the soils—is the subject of
an excellent general article by Robert B. Cook, "Man and the
Biogeochemical Cycles," <u>Environment</u> (September 1984).   The very
critical nitrogen cycle is discussed in C. C. Delwiche, "The
Nitrogen Cycle," <u>Scientific American</u> (September 1970).   It is
interesting to note that our fuels such as coal and petroleum can
be understood as "leakages" in the carbon cycle. For that view-
point see Guy Ourisson, Pierre Albrecht, and Michel Rohner, "The
Microbial Origin of Fossil Fuels," <u>Scientific American</u> (August
1984).

On the role that chemistry can play in agricultural development, L. W. Shemilt (ed.), <u>Chemistry and World Food Supplies</u>
(1983), is a compilation of papers by authors at the frontier of
research and is comprehensive, up-to-date, and useful for
understanding current and future problems and prospects.   Nyle C.
Brady's presentation at the Chemrawn conference, from which the
Shemilt book was drawn, was published as "Chemistry and World
Food Supplies," <u>Science</u> (November 26, 1982).   Brady was formerly
the director of the International Rice Research Institute and is
now an administrator with USAID concerned with science and
technology.   Brady also has an article in Richard Staples and
Ronald J. Kuhr (eds.), <u>Linking Research to Crop Production</u>
(1980), a book that does a good job on the problem defined by the
title.   This book is relevant to the inquiry of chemistry and
agriculture as well as to the other chapters on agriculture.
Another book with wide applicability on issues of science and
agriculture is Tom K. Scott (ed.), <u>Plant Regulation and World
Agriculture</u> (1979).   The National Research Council of the
National Academy of Sciences (U.S.) report, <u>World Food and
Nutrition Study</u> (1977), establishes a broad framework for
understanding the potential contribution of research to food
production, and the multivolume supporting papers offer a series
of more in-depth analysis of research possibilities.   One of
the areas of research with greatest potential is on nitrogen
fixation. A comprehensive work on the subject is J. R. Postgate,
<u>The Fundamentals of Nitrogen Fixation</u> (1982).   Recent research

and its significance is covered in J. R. Postgate, "New Kingdoms for Nitrogen Fixation," Nature (November 15, 1984).

CHAPTERS 9 AND 10

Too many of the discussions about trends in world development tend to be rich in theory and weak in data. There are many large and small data sources from the International Monetary Fund, the World Bank, the United Nations and many U.N. agencies such as Food and Agriculture Organization and International Labour Organization. One of the most useful data sources is the World Development Report, which has been published annually by the World Bank since 1978. Each report has a narrative section that either looks at a specific development problem or presents an overview of world economics. In each there is a statistical annex, World Development Indicators. Various U.N. agencies are following a similar format in publishing annual surveys and data sources on their particular concern, such as ILO's World Labour Report and UNICEF's The State of World Children. Most of these, such as the World Bank and FAO, have much larger, more comprehensive statistical compilations that are published annually or episodically. There are also regular strictly data sources, such as the Monthly Bulletin of Statistics of the United Nations and the International Financial Statistics (also monthly) of the IMF.

Good use of this data has been made by many authors. David Morawetz, Twenty-five Years of Economic Development 1950-1975 (1977) and Jacques Loup, Can the Third World Survive? (1982), are two of the best interpretive works on aggregate development trends since 1950. Unfortunately, most of their data ends in the mid to late 1970s, and it is increasingly necessary for them or someone else to bring this type of data analysis up to more recent years. A useful book is Pradip K. Ghosh (ed.), Population, Environment and Resources, and Third World Development (1984). It has a collection of analytical articles on world population, a substantial statistical section, a large annotated bibliography, and a directory of information sources. It is part of a series structured in the same format and called International Development Books. This series includes titles on appropriate technology, technology policy, energy policy, and other subjects directly related to our inquiry.

Many investigators have done for specific development issues what Morawetz and Loup did for overall development. There is an enormous wealth of fat books and articles on world population, as well as reports, magazines, and journals devoted to the subject. A short time in any good library would yield an abundance of research material on this important topic. A recent edition of a quality textbook on population is useful for background orientation. Two authors that I have found to be extremely good at pulling together and analyzing large quantities of population data are Nick Eberstadt, "Recent Declines in Fertility in Less Developed Countries and What Population Planners May Learn from Them," World Development (January 1980), and Ansley J. Coale, "Recent Trends in Fertility in Less Developed Countries,"

Science (August 26, 1983). On the relationship between income
and fertility, Robert Repetto, Economic Equality and Fertility
in Developing Countries (1979), is outstanding. The World
Development Report 1984 (1984) has a substantial comprehensive
analysis of world population and projections of future trends.
They are concerned about the recent slowing or reversal of some
population trends and of the not-so-favorable population trends
in Africa. A pessimistic, or at least concerned, interpretation
of this data is expressed by Robert McNamara, "Time Bomb or
Myth?" Foreign Affairs (Summer 1984). Reflections on Population
(1984), by Raphael Salas, executive director of the United
Nations Fund for Population Activities, is also an important book
to read on the subject. In an interesting dissenting position,
Julian Simon has gained a certain fame or notoriety in his
argument that more population means more creative geniuses and
that there is no evidence for population growth adversely
affecting economic development. Simon has written or edited a
number of books and articles on demography and development
trends, some of which we cite. A recent statement of his
position on population can be found in Julian Simon, "Myths of
Overpopulation," Wall Street Journal (August 3, 1984). Simon was
part of the United States delegation to the very important World
Population Conference in Mexico City in 1984 and was influential
in the U.S. delegation's position on issues. A good coverage of
the conference and Simon's ideas can be found in Fred Pearce, "In
Defense of Population Growth," New Scientist (August 9, 1984).
        John W. Mellor and Bruce F. Johnstone, in "The World Food
Equation," Journal of Economic Literature (June 1984), provide
an excellent overview of world food supply and its relationship
to other development objectives. Other superb overall works
on world food are Marilyn Chou and David Harmon, Jr., Critical
Food Issues of the Eighties (1979); Richard G. Woods (ed.),
Future Dimensions of World Food and Population (1981); C. Peter
Timmer, Walter P. Falcon, and Scott R. Pearson, Food Policy
Analysis (1983); Foreign Affairs and National Defense Division,
Congressional Research Service, Library of Congress, Feeding the
World's Population (1984); and D. Gale Johnson, "World Food and
Agriculture," and A. E. Harper, "Nutrition and Health in the
Changing Environment," both in Julian L. Simon and Herman Kahn
(eds.), The Resourceful Earth (1984). An excellent, up-to-date
collection of readings on agriculture is Carl K. Eicher and John
M. Staatz (eds.), Agricultural Development in the Third World
(1984). Clifford W. Lewis, "Global Food Security--A Manageable
Problem," Development Digest (July 1983), explains the mechanisms
that allow food deficient countries to continue to import minimal
food needs, even if they have severe balance-of-payment problems.
There is an excellent series of books on specific crops sponsored
by the International Agricultural Development Service. We used
one of them, Wheat in the Third World (1982), by Haldore Hanson,
Norman E. Borlaug, and R. Glen Anderson. Borlaug won a Nobel
Peace Prize for his research and development in wheat for what,
along with similar achievements in rice, is now popularly called

the green revolution. For those who have argued that the green revolution failed, an empirical refutation can be found in Grant M. Scobie, Investment in International Agricultural Research (1979). In a similar vein is "Economic Benefits from Research," Science (September 14, 1979), by Robert E. Evenson, Paul E. Waggoner, and Vernon W. Ruttan. Two of the most balanced assessments of the green revolution are Keith Griffin, The Green Revolution (1972), and Vernon W. Ruttan and Hans P. Binswanger, "Induced Innovation and the Green Revolution," in Binswanger and Ruttan, Induced Innovation (1978).

For the section on nutrition we relied on a limited number of sources. Miloslav Recheigl (ed.), Man, Food and Nutrition (1973), may be over a decade old in a rapidly changing field, but it is still useful and definitely not out of date. Three nutrition books recommended to me proved helpful. They are Sir Stanley Davidson et al., Human Nutrition and Dietetics (1979); Mary Alice Caliendo, Nutrition and the World Food Crisis (1979); and Eleanor Noss Whitney and Eva May Nunnelly Hamilton, Understanding Nutrition (1981).

On crop breeding we drew heavily on AID publications, Horizons and its previous incarnation, Agenda. This publication allows use of its material as long as it is properly cited, and we made use of this generosity. The articles used were John G. Blair, "Custom Tailored Crops," Agenda (July-August 1980), and Walt Rockwood, "'New' Biotechnology in International Agricultural Development," Horizons (November 1983). These articles are written by competent specialists so that they can be understood by anyone working in development. Another article that, along with those named above, forms an excellent introduction to the subject is Colin Tudge, "The Future of Crops," New Scientist (May 26, 1983). Two other recent articles on plant breeding are important, though they focus more closely on specific techniques or objectives. They are Lloyd W. Ream and Milton P. Gordon, "Crown Gall Disease and Prospects for Genetic Manipulation of Plants," Science (November 26, 1982), and Roger M. Gifford et al., "Crop Productivity and Photoassimulated Partitioning," Science (August 24, 1984).

There is a large literature on the benefits of agricultural research and the diffusion of new innovations. Many argue that agricultural research has the highest return of any investment, both as a domestic endeavor and a cooperative international undertaking. A simple way of getting to some of this literature is to start with some good recent books and follow their reference notes to other books and journal articles. No two books on the subject can be recommended more highly than Resource Allocation and Productivity in National and International Research (1977), by Thomas R. Arndt, Dana G. Dalrymble, and Vernon W. Ruttan (eds.), and Agricultural Research Policy (1982), by Vernon W. Ruttan.

The section on potential new crops cites a variety of popular articles as easy-to-read sources. Although many of them are well written and interesting, and some are even illustrated, the National Research Council of the National Academy of Sciences (U.S.) publication on each subject contains virtually all the essential information. Some of the popular pieces are written by or are interviews of people who were involved in the National Research Council studies. The two most important works are Underexploited Tropical Plants with Promising Economic Value (1975) and Tropical Legumes (1979). In addition, the National Academy of Sciences has published monographs on specific plants such as Guayule (1977), Leucaena (1977), The Winged Bean (1975), Amaranth (1984), and Jojoba (1984). Anyone even mildly serious about the subject should check the latest National Academy of Sciences publication list and the latest edition. A good recent survey of some crops with drylands potential (including jojoba and guayule) is C. Wiley Hinman, "New Crops for Arid Lands," Science (September 28, 1984), and an exploration of the possibility of salt-tolerant crops can be found in Jean L. Marx, "Plants: Can They Live in Salt Water and Like It," Science (December 7, 1979). Paul F. Knowles, Development of New Crops (October 1984), argues the need for new crops in the United States.

CHAPTERS 11 AND 12

The economic issue of the choice of technology has long been grounded in questions of the factors of production. The intensity of use of any factor of production--i.e., land, labor, capital (i.e., technology)--depends upon its relative scarcity or abundance (i.e., its price). Given the history of economic thinking on this subject, it would almost be axiomatic that in countries where the price of labor was low, economic choice would involve labor-intensive technologies.

In development economics this mode of thinking came to fruition in the 1950s. W. Arthur Lewis established the basic model in which a surplus of labor in agriculture could be used with inexpensive basic technology to create the capital to build a more advanced society. In productive enterprises wages would be at subsistence, profits would be high, and therefore capital would accumulate. As development progresses, wages would increase and more advanced technology would be used. See W. Arthur Lewis, "Economic Development with Unlimited Supplies of Labor," The Manchester School of Economic and Social Studies (May 1954). For a similar but more sophisticated theory, see John C. H. Fei and Gustav Ranis, Development of a Labor Surplus Economy (1964). Since 1964, Ranis has written extensively on Taiwan as the premier example of successful economic development, following the policy implicit in the Fei-Ranis model. Erich Schumacher's economic premises underlying his widely acclaimed book, Small is Beautiful (1973), are, in effect, an oversimplification of Lewis, Fei, and Ranis. Whatever one may think of the theoretical validity of Schumacher's work and others', such as Ivan Illich, Tools for Conviviality (1973), those stimulated by his thinking have done an enormous amount of good work in finding or creating appropriate technologies. In this regard, there seems to be an almost endless number of appropriate-technology magazines and journals, such as Vita News and Appropriate Technology. In his book, Transforming Traditional Agriculture (1965), Theodore W. Schultz questions whether there is a labor surplus in agriculture. If there were a labor surplus, meaning that the marginal product of labor was zero, then one would not expect population decline to lead to a fall in agricultural output. Schultz presents evidence that it does.

Vernon W. Ruttan and Yujiro Hayami have argued in books and articles that different factors of production give rise to different types of science and technology. In the United States, where land was abundant and labor relatively scarce, agricultural research and training was primarily in the area of labor-saving mechanical technology. In Japan in the late 19th century, with its extreme scarcity of land, biological research fostered increased yields per acre. Following this, they argue that the factor endowments of third world countries make continued biological research the most promising for their agricultural development. Among their singular and joint efforts, see Vernon W. Ruttan and Yujiro Hayami, Technology Transfer and Agricultural Development (1973).

For an excellent piece of research in agriculture, which, among the many things that it accomplishes, calls into question oversimplified theories of economic and technological change, see Derek Byerlee et al., Rural Employment in Tropical Africa (1977). The work represents a basic approach of factually studying each task and environment, then selecting, adapting, or designing the technology to fit that problem. It avoids rigid aprioristic theory, but that does not mean it lacks an empirically testable theory or theories of development.

The employment question in rural areas is a critical one. The same is true in urban areas. A seminal work on the employment issue is Francis Stewart and Paul P. Streeten, "Conflicts Between Output and Employment Objectives," in R. Robinson and P. Johnston (eds.), Prospects for Employment Opportunities in the 1970's (1972), reprinted in Richard Jolly et al., Third World Employment (1973). The technology and employment issue is covered in some detail in Frances Stewart, "Technology and Employment in LDC's," in Edgar O. Edwards, Employment in Developing Nations (1974), and Frances Stewart, Technology and Underdevelopment (1977). In "Unemployment and Economic Development," Economic Development and Cultural Change (October, 1984), Albert Berry and R. H. Sabot challenge the widely held thesis that urban unemployment has been increasing in developing countries. This is the best article on the subject that I have read in several years.

Often neglected in the theorizing on technology transfer or in the choosing of technology are concrete criteria for actual selection of technology. Richard Eckaus, in Technologies for Developing Countries (1977), has one of the best set of criteria for selecting appropriate technologies. W. Paul Strassman, in Technological Change and Economic Development (1965) and other works, provides reasoned arguments and criteria for selecting used machinery in manufacturing. World Development, September/ October 1977, is a special issue on "The Choice of Technology in Developing Countries"; it contains an excellent series of research articles on specific technologies and the possibilities (or lack thereof) of scaling down technologies to fit local factory endowments. One of the best books on technology for a specific region is James H. Street and Dilmus D. James (eds.), Technological Progress in Latin America (1979). Two excellent sets of articles are Appropriate Technology: Problems and Promises, edited by Nicolas Jequier (1976), and The Choice of Technology in Developing Countries (1975), by C. Peter Timmer et al. The International Labour Office is seeking to make the best integrated use of different kinds of technologies. This pragmatic and sensible activity is explored in a series of case studies, Blending of New and Traditional Technologies (1984), by A. Bhalla, D. James, and Y. Stevens (eds.). A short, solid, clear critique of alternate technologies can be found in Radhakrishna Rao, "When Alternatives are Inappropriate," New Scientists (1980). The intense interest in appropriate technology in the 1970s has lessened in the 1980s. Though the activists still have their publication, there has been a diminution of the type noted in this paragraph.

There are many different types and kinds of solar energy. What is called passive solar refers either to improved insulation or to various ways of trapping and conserving heat from the sun or in the ground. Prevention of heat loss also tends to reduce air exchange, allowing for the buildup of indoor pollutants. On indoor air pollution, Laurence S. Kirsch, "Behind Closed Doors," Harvard Environmental Law Review (1982), was about the best, most comprehensive and extensively documented article found. One need only read it and pursue its sources and other contemporary ones for a clear understanding of this issue. A recent popular account is Peter Grier, "After Working to Limit Air Pollution Outdoors, U.S. Looks Indoors," Christian Science Monitor (September 25, 1984). Another quality article is John P. Spengler and Ken Sexton, "Indoor Air Pollution," Science (July 1, 1983).

What is called active solar can involve anything from biomass conversion, to windmills or waterpower, to silicon cells for electricity generation. Biomass conversion generally uses either digestors to generate methane gas from animal and human manure and crop residue or the direct burning of firewood. This is generally the simplest technology, though it may involve plant genetics in developing high-yielding "energy crops." By far the very best book, in my judgment, on the range of biomass possibilities is Vaclav Smil, Biomass Energies (1983). It is technical but still readable, massively documented, judicious and balanced in its judgment, and comprehensive in its coverage. A smaller, slightly older but still outstanding book is Malcolm Slesser and Chris Lewis, Biological Energy Sources (1979). A book highly favorable to these energy sources is Russell E.

Anderson, Biological Paths to Self-Reliance (1979). One of the major concerns has been for the availability of energy for agriculture. A solid scientific study on the movement of energy through the ecosystem is David H. Miller, Energy at the Surface of the Earth (1981). This work is a necessary foundation for understanding use of the ecosystem for biomass energy generation. It is also relevant to the energy concerns explored in Chapter 7.

Silicon cells for converting the sun's rays into electricity can range from simple solar pumps, to solar cookers, to sophisticated arrays of solar cells coordinated by computer to capture sunlight. The manufacture of these cells and the improvements in this manufacturing to make them competitive is essentially what is called high technology, and there is a large technical body of literature on the subject. On the popular, if not cult, level, there is a sizeable array of pamphlets, books, and magazines such as Solar Age or Solar Times, many of which can be found at larger newsstands or libraries. For those interested in solar power, Robert Noyes' The Sun, Our Star (1982) not only gives a first-rate popular account of the sun as a star but also ends with a good account of solar power and its potential. It is a good book to begin one's study before going onto others more specifically directed to terrestrial solar energy use. A good history of solar energy is Ken Butti and John Perlin's A Golden Thread (1980). The authors note the continuous issue for direct solar energy of unimpeded access to the sun. This is an issue that continues to the present, as shown in Robert Kilborn, Jr., "Battling for Sunlight," Christian Science Monitor (January 23, 1982).

Solar energy as it is currently used would not include coal or petroleum, though the hydrocarbons originated in biomass. Renewability is a defining characteristic. For the hydrocarbons, we are seen to be drawing on stored energy capital. If we use biomass at a rate that can be regenerated, then for the solar advocates it is sustainable. Solar technologies, except the most sophisticated varieties, form the core of the Lovins's "soft energy." Of the many writings of Amory B. Lovins and Hunter Lovins on soft energy, none states the case better than Amory's article, "Energy Strategy," Foreign Affairs (October 1976), reprinted in Amory B. Lovins, Soft Energy Paths (1977). An overall perspective and advocacy of solar power can be found in Denis Hayes, Rays of Hope (1977), and Hazel Henderson, The Politics of the Solar Age (1981).

Many appropriate or small technologies may have substantial advantages in theory over existing technologies but still fail in practice. Until the causes of failure are better understood, attempts at improvements in traditional technologies, such as cookstoves, should not be rejected. Premature abandonment of a technology, be it "hard" or "soft," is as unwise as persistent use of a technology long after its obsolescence is demonstrated. The increased energy efficiency from enclosing an open fire for cooking with some kind of stove made with local materials is obvious. Much good work in research and field testing has

been done, but the overall results are not exciting. There
are some first-rate articles on the type of efforts and the
problems encountered, however: Jonathon B. Tucker, "Appropriate
Technology", Environment (April 1983); Bina Agarwal, "Diffusion
of Rural Innovation," World Development (April 1983); Paul
Harrison, "Appropriate Technology," New Scientist (November 20,
1980); and Anil Agarwal and Anita Anand, "Ask the Women Who Do
the Work," New Scientist (November 4, 1982). As these articles
show, sometimes the problems are technical; at other times they
are social or cultural.

Supporters of small technologies frequently use China as an
example of the successful use of their type of technology. No
one has written more or better on China's energy achievements, or
is cited more often on this subject, than Vaclav Smil. His
general article, "Renewable Energies," The Bulletin of the Atomic
Scientists (December 1979) is a good place to begin reading on
renewable energy or on energy in China. On technology in rural
China, two works stand out: Benedict Stavis, The Politics of
Agricultural Mechanization (1978), and Dwight Perkins, Rural
Small-Scale Industry in the People's Republic of China (1977).
Also worth reading is Jan Sigurdson, Rural Industrialization in
China (1977), and a later work by Sigurdson, Technology and
Science in the People's Republic of China (1980). More recently,
there is "Science and Technology in China," a special supplement
of The Bulletin of the Atomic Scientist (October 1984). On
urban technology, the work of Thomas G. Rawski is outstanding,
including China's Transition to Industrialization (1980). Other
works cited in the section on technology are uniformly good.

Even though China is attempting a transition to larger, more
sophisticated technologies, as the authors cited indicate, by
industrial-country standards it is still operating with a
predominance of small technologies. For those who consider small
technologies to be inherently benign, another fine book by Vaclav
Smil, The Bad Earth (1984), is a necessity. For the way in which
small size limited agricultural development throughout China's
history, see Francesca Bray, Science and Civilization in China,
vol. 4, part 2, Agriculture (1984). This is the latest issue in
a monumental set of volumes by Joseph Needham, the first volume
having been published in 1965. On the possibilities of more
sophisticated work on water utilization in China, studied in
relation to similar experience elsewhere, see Asit K. Biswas et
al. (eds.), Long Distance Water Transfer (1983).

Intellectual inquiry of any kind is a human endeavor and as
such it inevitably involves perceptual problems. It is easy to
understand how we might project our motivations on other forms of
life or our cultural values upon other peoples. An inquiry such
as ours in technology concerns human endeavor, but it also is
founded on physical properties. It is interesting to note, then,
that in the physical sciences there are concerns about an
"anthropic principle." For a popular article see George Gale,
"The Anthropic Principle," Scientific American (December 1981); a
longer, interesting, and very readable book describing the

conditions of the universe and giving a lucid account of the anthropic principle as interpretive theory, see P. C. W. Davies, The Accidental Universe (1982), and Davies' Superforce (1984), where he also discusses attempts for a "grand unified theory" of nature. For another nontechnical exposition in the context of cosmological inquiry in general, see Edward R. Harrison, Cosmology (1981). There is a body of technical writing, some of which is at best accessible to the nonspecialist only in intermittent prose sections. These are (in chronological order) R. H. Dicke, "Cosmology," Nature (November 4, 1961); C. B. Collins and S. W. Hawking, "Why is the Universe Isotropic?" Astrophysical Journal (March 1, 1973); B. J. Carr and M. J. Rees, "The Anthropic Principle and the Structure of the Physical Universe," Nature (April 12, 1979); and A. D. Linde, "The New Inflationary Universe Scenario," in G. W. Gibbons, S. W. Hawking, and S. T. C. Sikos (eds.), The Very Early Universe (1983), for a "weak anthropic principle."

CHAPTER 13

The histories of famine, disease, and natural catastrophes lend themselves to sensationalist writing. Some of these events, such as the Irish potato famine, have been of such consequence that they have attracted scholarly study and publication as well as the writing of quality popular books and articles. Writing on contemporary catastrophes tends to bring out the same variety of sensationalist writing, scholarly inquiry, and quality journalism. However, if the readers extract the verifiable data as to the actual numbers of human beings killed or maimed, then compare it to recent tragedies with similar historical experiences, they are virtually certain to arrive at a conclusion similar to Julian Simon's in "Resources, Population, Environment," Science (June 27, 1980). Both absolutely and proportionately, the impact of disasters in modern times is far less than was the experience of earlier times. The lower death and injury rates do not in the least lessen our determination to reduce these human tragedies even further. To the extent that we can argue that technology has been a major causal factor in these reductions, we can better understand the means to bring about further reductions.

On the political dimensions that turn a food shortage into a killing famine, see Amartya Sen, Poverty and Famines (1981), where he distinguishes between food availability and entitlement to it. Sen writes on famine in several areas but particularly in India. For India there is a sizeable body of literature on the history of famine (particularly in the 19th century) as part of the larger debate on the economic impact of British colonial rule and on agriculture, the green revolution, and development in the period since independence. Some of the better-known works are B. M. Bhatia, Famine in India (1963); Mohindar Singh Randhawa, Green Revolution (1974); C. H. Hanamantha Rao, Technological Change and Distribution of Gains in India (1975); and a more popular work, Kusam Nair, In Defense of the Irrational Peasant (1979). This

chapter cites statistics on deaths from catastrophes in Europe
and Asia (India and China) and to a lesser extent in North
America. Similar types of data may be found for Latin America in
Nicolas Sanchez-Albornez, The Population of Latin America (1974).
     Siegfried Giedion's Mechanization Takes Command (1948) has
long been a standard work on the history of household technology.
Interesting though these sections of the book are, they are
aseptic in tone and the household technology seems more like
something in a mail-order catalog setting than a household where
people actually lived. A lighter, delightfully illustrated book
in a similar vein is Harvey Green, The Light of the Home (1983).
The book has a superb chapter on "the tyranny of housework."
Reay Tannahill, Food in History (1973), has some excellent
chapters on the impact of science and the Industrial Revolution
on the quality and availability of food in the household.
Throughout this book (particularly in Chapter 12) I have shown
that technology as a problem-solving process is a bettering
activity, not a perfecting one. Thus, our household food
supply is better and safer than it has ever been, a point
strongly made by A. E. Harper, "Nutrition and Health in the
Changing Environment," in Julian Simon and Herman Kahn (eds.).
The Resourceful Earth (1984). Though we may have virtually
eliminated botulism and most life-threatening food contaminants,
in the process we may have bred potentially more lethal strains
of salmonella. See Fred Pearce and Jeremy Cherfas, "Antibiotics
Breed Lethal Food Poisons," New Scientist (September 13, 1984).
The reports upon which the Pearce and Cherfas article, as well
as several newspaper accounts, are based can be found in "Animal-
to-Man Transmission of Antimicrobials Resistant Salmonella:
Investigations of U.S. Outbreaks, 1971-1983," Science (August 24,
1984), by Scott D. Holmberg, Joy G. Wells and Mitchell L. Cohen,
and "Drug Resistant Salmonella from Animals Fed Antimicrobials,"
The New England Journal of Medicine (September 6, 1984), by Scott
D. Holmberg et al. Two news articles on the issue are Marjorie
Sun, "In Search of Salmonella's Smoking Gun," Science (October 5,
1984), and Marjorie Sun, "Use of Antibiotics in Animal Feed
Challenged," Science (October 12, 1984). One author highly
critical of the reports and news articles on them is Stephen
Budiansky, "Jumping the Smoking Gun," Nature (October 4, 1984).
Problem solving can and often does create new problems. All
our theory and inquiry argues is that technological "progress"
solves greater problems than it creates. The evidence we present
shows that the improvements in aggregate have been substantial
and dramatic, whether it be in food supply, shelter, or life
expectancy.
     Two more recent books on households and technology are Susan
Strasser, Never Done (1982), and Ruth Schwartz Cowan, More Work
for Mother (1983). Never Done is a masterful work in which
Strasser uses factual description to give the reader a heightened
sense, a "feel," for what it was like to live in an American
household over the past two centuries or more. She is particularly good in describing indoor life in cold winter weather.

In fact, her writing is so vivid that she is not believable
when she tries to minimize the degree to which these changes
reflect an overall pattern of improvement. An excellent
complement, if not corrective, to Strasser is Naomi Bliven's
review of her book, "Home, Sweet Home," New Yorker (September 6,
1982). Bliven argues that "though Miss Strasser philosophizes
poorly, she has researched magnificently," an assessment with
which I wholeheartedly concur.

More Work for Mother is less vivid in its description and
more technical and theoretical in its analysis. Cowan looks at
household technology as part of a larger system of technologies.
She maintains that the most startling transformations in
household technologies have been for those tasks mainly done by
men. She argues, correctly, that this pattern of change was not
the result of any technological imperatives and sketches out
what were alternate possible pathways of change. Despite the
strengths of this thought-provoking book, Cowan's arguments are
forced and unconvincing when she tries to argue that technology
has not lessened woman's household burdens.

When we turn to an historical look at family life, one work
stands out, Lawrence Stone, The Family, Sex and Marriage in
England 1500-1800 (1977). For the United States, a growing
number of sociologists are questioning what some call the myth of
the decline of the American family. An excellent, clearly
written statement of this position is Mary Jo Bane, Here to Stay
(1976). Two excellent collections on the family are Arlene
Skolnick and Jerome H. Sholnick (eds.), Family in Transition
(1980), and Tamara K. Hareven (ed.), Transitions (1978). The
Hareven book contains an essay, "Changing Configurations of the
Life Course," by Peter Uhlenberg, whose work here and elsewhere
is a tour de force on family statistics.

A different perspective on changes in family life than the
one we present can be found in Peter Laslett, The World We Have
Lost (editions 1965, 1971, 1984), and Christopher Lasch, Haven in
a Heartless World (1979).

One of the most famous studies of a community and its family
life is the work of Robert S. Lynd and Helen Merrel Lynd on
Muncie, Indiana. The first of their series of books was the
classic Middletown (1929). Their work forms the baseline for
current research in a restudy that has resulted in Middletown
Families (1982), by Theodore Caplow et al. A careful reading of
both books, especially the conclusions of Middletown Families,
clearly does not support a thesis of a decline in family life.
In fact, the evidence points in the other direction.

Philippe Aries in many ways initiated the modern study of
the history of childhood with his book Centuries of Childhood
(1962). Important though his work is, it is not definitive in
that many recent writers have voiced their dissent. On the
history of infanticide, the most scholarly author is William L.
Langer in such works as "Infanticide," History of Childhood
Quarterly (Winter 1974). Another good source is Lloyd de
Mause (ed.), The History of Childhood (1974). Unfortunately,

de Mause's writing too often has a sensationalist style of presentation that may make the serious reader question the validity of his data. Today, in very poor countries, some of the problems for children remain, though they are more ones of neglect or nutritional deprivation. Cultural bias tends to make girl babies the more likely victims. See the fine study by Barbara Miller, The Endangered Sex (1981). Females have long been the chief victims of brutality, be they infant or older. See Edward Shorter, A History of Women's Bodies (1982).

Infanticide (1984) by Glenn Hausfater and Sarah Blaffer Hrdy (eds.) is the best work on the subject that I have read. Most of the studies in the book concern infanticide in a number of different animals. The essays on infanticide among humans are outstanding. The book's massive bibliography is one of the most comprehensive in the literature on infanticide.

Paul Ehrlich, The Population Bomb (1971), writes of the "brutal and heartless decision" that population growth will force upon us. The argument in the last three chapters of my book is that lack of technology creates the necessity for these cruel choices. Many historians now argue that our modern sensibilities (at least in Western culture) about the worth of "nature" and its preservation, about the rights of other living creatures, and even about our attitudes against cruelty and our sympathy for the rights of other human beings, have developed largely in urban areas in the past few centuries. No one has done a more erudite, thought-provoking work on this subject than the distinguished British historian, Keith Thomas, in Man and the Natural World (1983). Even those who presumed to love nature didn't treat it as we would expect today. From Thomas, we learn that "in the eighteenth century the first impulse of many naturalists on seeing a rare bird was to shoot it." Most of the famous and great bird books were written and illustrated by those who went out and killed the finest specimens so that they could draw them, a point made by Robert Welker, Birds and Men (1955). It wasn't until well into the nineteenth century, with the technology of photographs, that bird books could be illustrated without first killing the birds.

The first chapter of this book was originally devised in response to arguments made by Kirkpatrick Sale, in Human Scale (1980), and by others that modern technology is contradictory to the life process as we know it. We close with the main theme of Sale's book, that smallness is a biological virtue applicable to technology. Sale claims to have derived his thesis from J. B. S. Haldane's essay, "On Being the Right Size," in Haldane, Possible Worlds (1928), yet Haldane found advantages to smallness and to bigness. Another author who sees possible advantages to larger size because of the greater potential use of energy is William Day, Genesis on Planet Earth (1984, 2nd ed.). Thomas A. McMahon and John Tyler Bonner, On Size and Life (1983), find varied advantages and disadvantages to differing sizes. They find that small life forms would not be able to use fire or tools and consequently could not create civilization as we know it, or technology even of the appropriate kind advocated by Sale. Study of the significance of size in life forms has increased in recent years. Two new works are Knut Schmidt-Nielson, Scaling (1984), and William A. Calder III, Size, Function, and Life History (1984). Another scholarly study uses body size as a measurable component in quantitative inquiry in ecology. In this research by Robert Henry Peters, The Ecological Implications of Body Size (1983), where the central purpose is to develop a quantitative scientific model, one searches in vain for the supposed self-evident truth as to the biological virtues of smallness.

In fact, despite whatever sensation Sale may have generated with his book, I find no evidence of any scientific interest in his thesis. Like many of the more grandiose claims of the appropriate-technology movement, the assertions are more eloquent than verifiable, more utopian than scientific; and as time passes, the ideas are seen to be more ephemeral than sustainable. Appropriate technology was touted in the 1970s as an idea whose time had come; it is now one whose time has passed. What is needed is an understanding of the continuing appropriateness of the technological process to the human endeavor. It is hoped that A Theory of Technology serves this purpose.

REFERENCES

Abelson, Philip H. 1982. Improvement of grain crops. Science
216, no. 4543. April 16.

Abelson, Philip H., and Allen L. Hammond, eds. 1976. Materials:
Renewable and nonrenewable resources. Washington, DC:
American Association for the Advancement of Science.

Agarwal, Anil, and Anita Anand. 1982. Ask the women who do the
work. New Scientist 96, no. 1330. November 4.

Agarwal, Bina. 1983. Diffusion of rural innovations: Some
analytical issues of wood burning stoves. World Development
11, no. 4. April.

Ambroggi, Robert P. 1980. Water. Scientific American 243, no.
1. September.

Ames, Bruce N. 1983. Dietary carcinogens and anticarcinogens:
Oxygen radicals and degenerative diseases. Science 221, no.
4617. September 23.

Anderson, Russell E. 1979. Biological paths to self-reliance:
A guide to biological solar energy conversion. New York:
Van Nostrand Reinhold Co.

Aries, Philippe. 1962. Centuries of childhood: A social
history of family life. New York: Knopf.

Arndt, Thomas K., Dana G. Dalrymple, and Vernon W. Ruttan, eds.
1977. Resource allocation and productivity in national and
international agriculture. Minneapolis: Univ. of Minnesota
Press.

Ash, Caroline, David Crompton, and Anne Keymer. 1984. Nature's
unfair food tax. New Scientist 101, no. 1400. March 8.

Asimov, Isaac. 1979a. Bacterial engineering. Realities, no. 7.
September–October.

———. 1979b. A choice of catastrophes. New York: Simon &
Schuster.

Atkins, Peter William. 1984. The second law. New York:
Scientific American Library.

Ayres, C. E. 1944. The theory of economic progress: A study of
the fundamentals of economic development and cultural
change. Chapel Hill: Univ. of North Carolina Press.

———. 1962. Towards a reasonable society. Austin, Texas: Univ.
of Texas Press.

Aziz, Mahmoud. 1979. 'Tree' of palent-leucaena in the

Philippines. The IDRC Reports 8, no. 4. December.

Bach, Wilfred, ed. 1983a. Carbon dioxide: Current views and developments in energy/climate research. Dordrecht, Netherlands: D. Reidel Publishing.

---. 1983b. Our threatened climate: Ways of averting the CO_2 problem through rational energy use. Dordrecht, Netherlands: D. Reidel Publishing.

Bach, Wilfred, Jürgen Tankrath, and William Kellogg, eds. 1979. Man's impact on climate. Proceedings of an international conference held in Berlin, June 14-16, 1978. Amsterdam, Netherlands: Elsevier.

Baden, John, ed. 1984. The vanishing farmland crisis: Critical views of the movement to preserve agricultural land. Lawrence, Kansas: Univ. Press of Kansas for the Political Economy Research Center, Bozeman, Montana.

Bajracharya, Deepak. 1983. Fuel, food or forest? Dilemmas in a Nepali village. World Development 11, no. 12. December.

Bane, Mary Jo. 1976. Here to stay: American families in the twentieth century. New York: Basic Books.

Barbour, Ian, Harvey Brooks, Sanford LaKoff, and John Opie. 1982. Energy and American values. New York: Praeger.

Barnett, Harold J., and Chandler Morse. 1963. Scarcity and growth: The economics of natural resource availability. Baltimore: Johns Hopkins.

Barnett, Harold M., Gerald M. Van Muiswinkel, Mordecai Sheckter, and John G. Myers. 1984. Global trends in non-fuel minerals. In Simon and Kahn.

Barney, Gerald O., ed. 1980. The global 2000 report to the President; vol. 3, Entering the twenty-first century. Washington, DC: USGPO.

Bates, Marston. 1967. Gluttons and libertines: Human problems of being natural. New York: Vintage Books.

Beardsley, Tim. 1983. Animals as gamblers. New Scientist 98, no. 1360. June 2.

Beckman, Petr. 1984. Solar energy and other 'alternative' energy sources. In Simon and Kahn.

Bell, John. 1984. The ceramics age dawns. New Scientist 101, no. 1394. January 26.

Benge, Michael. 1983. Miracle tree: Reality or myth? Horizons (Agency for International Development) 2, no. 6. June.

Berg, Alan. 1978. The nutrition factor: Its role in nutritional development. Washington, DC: Brookings Institution.

Bernstein, Jeremy. 1979. Profiles--Hans Bethe, part 2. The New Yorker 60, no. 43. December 16.

Berry, Albert, and R. H. Sabot. 1984. Unemployment and economic development. Economic development and cultural change 33, no. 1, October.

Bettman, Otto L. 1974. The good old days--they were terrible. New York: Random House.

Bhalla, A., D. James, and Y. Stevens. 1984. Blending of new and traditional technologies: Case studies. Dublin, Ireland:

Tycooly International Publishing Limited.

Bhatia, B. M. 1963. Famines in India: A study in some aspects of the economic history of India (1860-1965). Bombay: Asia Publishing House.

Billington, David P. 1983. The tower and the bridge: The new art of structural engineering. New York: Basic Books.

Billington, David P., and Robert Mark. 1984. The cathedral and the bridge: Structure and symbol. Technology and Culture 25, no. 1. January.

Binswanger, Hans, and Vernon W. Ruttan. 1978. Induced innovation: Technology, institutions and development. Baltimore: Johns Hopkins.

Biswas, Asit K., Zuo Dakang, James E. Nickum, and Liu Changming, eds. 1983. Long-distance water transfer: A Chinese case study and international experiences. Dublin, Ireland: Tycooly International Publishing Limited for the United Nations University, Water Resources Series, no. 3.

Blair, John G. 1980. Custom tailored crops. Agenda (Agency for International Development) 3, no. 6. July-August.

Bliven, Naomi. 1982. Home, sweet home. Review of Never done, by Susan Strasser. New Yorker. September 6.

Boffey, Philip. 1984. Plans to release new organisms into nature spur concern. New York Times. June 12.

Böhm-Bawerk, Eugen V. 1930. The positive theory of capital. Trans. by William Smart. London: Macmillan.

---. 1957. Capital and interest: A critical history of economical theory. Trans. by William Smart. (Original English trans. 1890.) London: Frank Cass & Co.

Borlaug, Norman. 1981. Using plants to meet world food needs. In Woods.

Boserup, Esther. 1965. The conditions of agricultural growth: The economics of agrarian change under population pressure. Chicago: Aldine.

Bousquet, Joseph. 1983. Food and cultivation potential of morama beans from the Kalahari Desert Island. Paper presented at the American Association for the Advancement of Science. Detroit. May.

Bossong, Ken. 1979. Hazards of solar energy. Washington, DC: Citizens Energy Report, Project Series no. 41.

Boyer, J. S. 1982. Plant productivity and environment. Science 218, no. 4571. October 29.

Brace, C. Loring. 1967. The stages of human evolution: Human and cultural origins. Englewood Cliffs, NJ: Prentice Hall.

Brady, Nyle C. 1980. The evaluation and removal of constraints to crop production. In Staples and Kuhr.

---. 1982. Chemistry and world food supplies. Science 218, no. 4575. November 26.

Braidwood, Robert J. 1964. Prehistoric men. Glenview, IL: Scott, Foresman.

Braudel, Fernand. 1973. Capitalism and material life 1400-1800. New York: Harper & Row.

Braun, Ernst. 1984. Wayward technology. Westport, CT: Greenwood.

Bray, Francesca. 1984. Agriculture 6, pt. 2. In Needham.

Bremmer, David, and Ann Prescott. 1984. Painting with light. New Scientist 102, no. 1411. May 24.

Bressani, R., and L. G. Elias. 1973. Development of new highly nutritious food. In Recheigl.

Brill, Winston J. 1977. Biological nitrogen fixation. Scientific American 236, no. 3. March.

Broad, William J. 1984. Does genius or technology rule science? New York Times. August 7.

Brobst, Donald. 1978. Fundamental concepts for the analyses of resource availability. In Smith.

Brody, Jane. 1979. How wholesome is 'natural' food? New York Times. December 12.

———. 1984. Ancient, forgotten plant, now 'grain of the future.' New York Times. October 16.

Bronowski, Jacob. 1956. Science and human values. New York: Harper & Row.

Brooks, David B., and Peter W. Andrews. (1974) 1976. Mineral resources, economic growth, and world population. Science 185, no. 4145. July 5. Reprinted in Abelson and Hammond.

Brooks, Harvey. 1980. Technology, evolution, and purpose. Daedalus: Journal of the American Academy of Arts and Sciences 109, no. 1. Winter.

Brown, Jonathon. 1980. In detail: Velasquez's Las Meninas. Portfolio: The Magazine of Visual Arts 2, no. 1. February-March.

Brown, Lester R. 1974. By bread alone. New York: Praeger.

———. 1978. The twenty-ninth day: Accommodating human needs and numbers to the earth's resources. New York: W. W. Norton.

Brown, Lester, William Chandler, Christopher Flavin, Sandra Postel, Linda Starke, and Edward Wolf. 1984. State of the world 1984. New York: W. W. Norton.

Brown, Lester, and Edward Wolf. 1984. Soil erosion: Quiet crises in the world economy. Washington, DC: Worldwatch Institute.

Brown, Peter. 1967. The later Roman empire. The Economic History Review. 2nd series, vol. 20, no. 2. August.

Bruce-Briggs, B. 1975. The war against the automobile. New York: E. P. Dutton.

Bryant, John. 1980. Systems theory, survival, and the back-to-nature movement. Technology and Culture 21, no. 2. April.

Brynka, Barbara. 1983. Genes on the cob. Science Digest 91, no. 4. April.

Buckley, Shawn. 1979. Sun up to sun down. New York: McGraw-Hill.

Budiansky, Stephen. 1984a. Anatomy of a pressure group. Nature 309, no. 5966. May 24.

———. 1984b. Jumping the smoking gun. Nature 311, no. 5986. October 4.

Bulletin of the Atomic Scientists. 1984. Science and technology in China. Special supplement in vol. 40, no. 8. October.

Bureau of Public Affairs, U.S. Department of State. 1979. The planetary product: Progress despite the blues 1977–1978. Special Report no. 58. Washington, DC: U.S. State Dept.

Burgess, Anthony. 1965. Language made plain. New York: Thomas Y. Crowell.

Buringh, P., H. D. J. van Heemst, and G. J. Staring. 1975. Computation of the absolute maximum food production of the world. Wageningen, The Netherlands: Dept. of Tropical Soil Science, Agricultural University.

Butti, Ken, and John Perlin. 1980. A golden thread: 2500 years of solar architecture and technology. New York: Cheshire/ Van Nostrand/Reinhold Books.

Byerlee, Derek, Carl K. Eicher, Carl Liedholm, and Dunstant S. C. Spencer. 1977. Rural employment in tropical Africa: Summary of findings. East Lansing, Michigan: African Rural Economy Program, Dept. of Agricultural Economics, Michigan State Univ., working paper no. 20. February.

Caiden, Naomi, and Aaron Wildavsky. 1974. Planning and budgeting in poor countries. New York: John Wiley & Sons.

Calder, William A. III. 1984. Size, function, and life history. Cambridge, MA: Harvard Univ. Press.

Calef, Charles F. 1976. An environmental critique of solar power by bioconversion methods. (BNL-21955) Upton, NY: Brookhaven National Laboratory.

Caliendo, Mary Alice. 1979. Nutrition and the world food crisis. New York: Macmillan.

Calvin, Melvin. 1981. Chemical evolution. In Skinner.

Campbell, Bernard G., ed. 1982. Humankind emerging. Boston: Little, Brown. Reprinted by permission. Copyright © 1982. Little, Brown, & Co.

Caplow, Theodore, Howard M. Bahr, Bruce A. Chadwick, Reuben Hill, and Margaret Holmes Williamson. 1982. Middletown families: Fifty years of change and continuity. Minneapolis: Univ. of Minneapolis Press.

Carman, John S. 1979. Obstacles to mineral development. New York: Pergamon.

Carola, Robert. 1981. Sunlight to sugar. In Ripley.

Carr, B. J., and M. J. Rees. 1979. The anthropic principle and the structure of the physical universe. Nature 278, no. 5705. April 12.

Castle, Douglas M. 1980. Keep regulations that work. Houston Chronicle. December 19.

Census Office. 1903. Statistical atlas of the United States 1900. Washington, DC: U.S. Census Office.

Chaisson, Eric. 1981. Cosmic dawn: The origins of matter and life. Boston: Little, Brown & Co.

Chapman, David J., and J. William Schopf. 1983. Biological and biochemical effects of an aerobic environment. In Schopf.

Chaudhuri, Pramet. 1979. The Indians economy: Poverty and development. London: Crosby Lockwood Staples.

Cherfas, Jeremy. 1984a. The difficulties of Darwinism. New

Scientist 102, no. 1410. May 17.

———. 1984b. Test-tube babies in zoos. New Scientist 104, no. 1433, December 6.

Chilton, Mary-Dell. 1983. A vector for introducing new genes into plants. Scientific American 248, no. 6. June.

Chou, Marylin. 1979. The preoccupation with food safety. In Chou and Harmon.

Chou, Marylin, and David Harmon, Jr. 1979. Critical food issues of the eighties. New York: Pergamon Press.

Christiansen, M. N. 1982. World environmental limitations to food and fiber cultures. In Christiansen and Lewis.

Christiansen, M. N., and Charles F. Lewis, eds. 1982. Breeding plants for less favorable environments. New York: John Wiley & Sons.

Ciferri, Orio. 1981. Let them eat algae. New Scientist 91, no. 1272. September 24.

Chynoweth, A. G. 1976. Electronic materials: Functional substitutes. Science 191, no. 4228. February 20. Reprinted in Abelson and Hammond.

Claiborne, Craig. 1984. Food historian helps preserve native cuisines. New York Times. July 11.

Clark, David H. 1984. Superstars: How stellar explosions shape the destiny of our universe. New York: McGraw-Hill.

Clark, Graham, and Stuart Piggott. 1965. Prehistoric societies. New York: Alfred A. Knopf.

Clark, William C. 1982. Carbon dioxide review: 1982. Oxford: Clarendon Press.

Clayton, William Robert, Jr. 1980. Overshoot: The ecological basis of revolutionary change. Urbana: Univ. of Illinois Press.

Cloud, Preston. 1981. Evolution of eco-systems. In Skinner.

Coale, Ansley J. 1983. Recent trends in fertility in less developed countries. Science 221, no. 4613. August 26.

Cohen, Mark Nathan. 1977. The food crisis in prehistory: Overpopulation and the origins of agriculture. New Haven, CT: Yale Univ. Press.

Cole, H. S. D., Christopher Freeman, Marie Jahoda, and K. L. R. Pavitt, eds. 1974. Thinking about the future: A critique of the limits to growth. London: Chatto & Windus, for Sussex Univ. Press. (U.S. reprint 1974, Models of doom: A critique of the limits to growth. New York: Universe Books.)

Collins, C. B. and S. W. Hawking. 1973. Why is the universe isotropic? Astrophysical Journal 180, pt. 1, no. 2. March 1.

Consumer Reports. 1980. It's natural: It's organical, or is it? Vol. 45, no. 7. July.

Cook, Robert B. 1984. Man and the biogeochemical cycles—interacting with the elements. Environment 26, no. 7. September.

Cornea, Giovanni Andrea. 1984. A survey of cross-sectional and time-series literature affecting child welfare. Special issue of World development, the impact of world recession on children. Ed. by Richard Jolly and Giovanni Andrea Cornea, vol. 12, no. 3. March.

Cornell, James. 1976. The great international disaster book. New York: Charles Scribner's & Son.

Cornell, John. 1984. Analogy and technology in Darwin's vision of nature. Journal of the History of Biology 17, no. 3. Fall.

Coulter, John K. 1980. Crop improvement. In Staples and Kuhr.

Courrier, Kathleen, publications director. 1984. The global possible: Resources, development, and the new century. The statement and action agenda of an international conference by the World Resources Institute. New York: World Resources Institute.

Cowan, Ruth Schwartz. 1983. More work for mother: The ironies of household technology from the open hearth to the microwave. New York: Basic Books.

Cox, Meg. 1979. Plant magic. The Wall Street Journal. October 10.

Creel, H. G. 1965. The role of the horse in Chinese history. The American Historical Review 70, no. 3. April.

Cresswell, David R. 1980. Extraterrestrial materials processing and construction. Houston, Texas: Lunar and Planetary Institute, National Aeronautics and Space Administration, Lyndon B. Johnson Space Center. January 31.

Crosson, Pierre R., ed. 1982. The cropland crisis: Myth or reality? Baltimore: The Johns Hopkins Univ. Press for Resources for the Future.

———. 1984. Agricultural land: Will there be enough? Environment 26, no. 7. September.

Cuny, Frederick C. 1983. Disasters and development. New York: Oxford Univ. Press.

Daly, John. 1984. New uses for nature's treasures. Horizons 3, no. 4. Fall.

Daumas, Maurice. 1969. A history of technology and invention: Progress through the ages; vol. 1, The origins of technological civilization. Trans. by Eileen B. Hennessy. New York: Crown.

Davidson, Sir Stanley, R. Passmore, J. F. Brock, and A. S. Truswell. 1979. Human nutrition and dietetics, 7th ed. Edinburgh: Churchill and Livingston.

Davies, P. C. W. 1982. The accidental universe. Cambridge: Cambridge Univ. Press.

———. 1984. Superforce: The search for a grand unified theory of nature. New York: Simon and Schuster.

Davis, Shelton H. 1977. Victims of the miracle: Development and the Indians of Brazil. Cambridge: Cambridge Univ. Press.

Davis, Shelton H., and Robert Mathews. 1972. The geological imperative: Anthropology and development in the Amazon basin of South America. Cambridge, Mass.: Anthropology Resources Center.

Day, William. 1984. Genesis on planet Earth: The search for life's beginning. New Haven: Yale Univ. Press.

Dearborn, Ned W. 1980. Nonfuel minerals projections. In Barney.

Decker, Peter. 1981. Coupling to solar energy: Sensitized photoreactions—the primary source of self organization. In Wolman.

Deevey, E. S., Jr. 1970. Mineral cycles. Scientific American 223, no. 3. September.

Deffeyes, Kenneth S., and Ian D. MacGregor. 1980. World uranium resources. Scientific American 292, no. 1. January.

Deggens, Egon T., S. Kempe, and V. Ittekot. 1984. Monitoring carbon in world rivers. Environment 26, no. 9. November.

DeGregori, Thomas R. 1969. Technology and the economic development of the tropical African frontier. Cleveland: Case Western Reserve Univ. Press.

———. 1974. Technology and economic change: Essays and inquiries. Comox, BC: Peter McLoughlin Associates.

———. 1980. Instrumental criteria for assessing technology: An affirmation by way of a reply. Journal of Economic Issues 14, no. 1. March.

———. 1982. The back-to-nature movement: Alternative technologies and the inversion of reality. Technology and Culture 23, no. 2. April.

———. 1984. Cancer and testing. Houston Chronicle. August 5.

DeGregori, Thomas R. and Oriol Pi-Sunyer. 1966. Technology, traditionalism and military establishments. Technology and Culture 7, no. 3. Summer.

Delwiche, C. C. 1970. The nitrogen cycle. Scientific American 223, no. 3. September.

de Mause, Lloyd, ed. 1974. The history of childhood. New York: The Psychohistory Press.

de Solla Price, Derek J. 1984. Of sealing wax and string. Natural History 93, no. 1. January.

Devine, T. E. 1982. Genetic fitting of crops to problem soils. In Christiansen and Lewis.

Dewey, John. 1929. The quest for certainty: A study of the relation of knowledge and action. New York: Minton, Balch & Co.

De Young, Garrett. 1983. Crop genetics: The seeds of revolution. High Technology 3, no. 6. June.

Dicke, R. H. 1961. Cosmology: Dirac's cosmology and Mach's principle. Nature 192, no. 4801. November 4.

Dickson, David. 1983. UNIDO hopes for biotechnology center. Science 221, no. 4618.

Dorfman, Ariel. 1984. Bread and burnt rice: Culture and economic survival in Latin America. Grassroots Development: Journal of the Inter-American Foundation 8, no. 2.

Dubos, Rene. 1980. The wooing of the Earth. New York: Charles Scribner's & Sons.

Dubos, Rene, and Jean Paul Escande. 1980. Quest: Reflections on medicine, science and humanity. Trans. by Patricia Ramum. New York: Harcourt, Brace, Jovanovich.

Duffy, M. C. 1983. Mechanics, thermodynamics and locomotive design: The machine-ensemble and the development of industrial dynamics. History and Technology 1, no. 1.

Duke, James A. 1982. Plant germ plasm resources for breeding of crops adapted to marginal environments. In Christiansen and Lewis.

Dunkerly, Joy, ed. 1980. International energy strategies: Proceedings of the 1974 IAEE/RFF conference. Cambridge, MA: Oelgeschlager, Gunn and Hain.

Duvick, David N. 1983. Improved conventional strategies and methods for selection and utilization of germ plasm. In Shemilt.

Dworkin, Judith. 1974. Global trends in natural disasters, 1947-1973. Natural Hazard Working Paper no. 26. Boulder Institute of Behavioral Science, Univ. of Colorado.

Eberstadt, Nick. 1980. Recent declines in fertility in less developed countries and what population planners may learn from them. World Development 8, no. 1. January.

Ebinger, Charles K. 1984. Eclipse of solar power leaves a burning need. Wall Street Journal. October 29.

Eckaus, Richard. 1977. Technologies for developing countries. Washington, DC: National Academy of Sciences.

Eckholm, Erik P. 1976. Losing ground: Environmental stress and world food prospects. New York: Norton.

---. 1984a. Fatal natural disasters on rise. New York Times. July 31.

---. 1984b. Theory on man's origins challenged. New York Times. September 4.

The Economist. 1980. Riches from the mobile Earth. Vol. 272, no. 7146. August 16.

---. 1982. Bringing paintings back to life. December 25.

---. 1984. Agrobacteria: Trojan bug. Vol. 293, no. 7363. October 13.

Edwards, Edgar O., ed. 1974. Employment in developing nations: Report on a Ford Foundation study. New York: Columbia Univ. Press.

Efron, Edith. 1984. The apocalyptics, cancer, and the big lie: How environmental politics controls what we know about cancer. New York: Simon and Schuster.

Ehrlich, Paul R. 1968a. The coming famine. Natural History 77, no. 5. May.

---. 1968b. Review of The limits to altruism, by Garrett Hardin. Human Nature 1, no. 3. March.

---. 1968c. The countdown to disaster. Wall Street Journal. December 3.

---. 1969a. Eco-catastrophe. Ramparts 8, no. 3. September. Reprinted in Current 3. October.

---. 1969b. World population: Is the battle lost? Reader's Digest 94, no. 2. February.

---. 1971. The population bomb. Rev. ed. New York: Ballantine.

Ehrlich, Paul R., and Anne H. Ehrlich. 1971. What happened to the population bomb? Human Nature 2, no. 1. January.

Eicher, Carl K., and John M. Staatz, eds. 1984. Agricultural development in the third world. Baltimore: The John Hopkins Univ. Press.

Eisenbud, Merril. 1978. Environment, technology, and health: Human ecology in historical perspective. New York: New York Univ. Press.

Eisenstein, Elizabeth L. 1979. The printing press as an agent of change. Cambridge: Cambridge Univ. Press. Reprinted by permission. Copyright © 1979.

---. 1983. The printing revolution in early modern Europe. Cambridge: Cambridge Univ. Press.

Engel, Peter. 1984. Against the currents of chaos. The Sciences 24, no. 5. September/October.

Environmental Protection Agency. 1983. Can we delay a greenhouse warming? Washington, DC: U.S. Environmental Protection Agency.

Etienne, Gilbert. (1977) 1978. Foodgrain production and population in Asia: China, India and Bangladesh. World Development 5, nos. 5-7. Reprinted in Sinha.

Evans, L. T. 1980. The natural history of crop yield. American Scientist 68, no. 4. July-August.

Evenson, Robert E., Paul E. Waggoner, and Vernon W. Ruttan. 1979. Economic benefits from research: An example from agriculture. Science 205, no. 4411. September 14.

Fagan, Brian M. 1980. People of the Earth. Boston: Little, Brown.

Fairbank, John K. 1979. The new two China problem. New York Review of Books 36, no. 2. March 8.

Farnsworth, Clyde H. 1984. A doubling of the world's population to 10 billion seen by year 2050. New York Times. July 11.

Feather, Frank, ed. 1980. Through the '80s: Thinking globally, acting locally. Washington, DC: World Future Society.

Fei, John C., and Gustav Ranis. 1964. Development of a labor surplus economy: Theory and policy. Homewood, IL: Richard D. Irwin.

Feibleman, James K. 1965. Artifactualism. Philosophy and Phenomenological Research 25, no. 4. June.

---. 1966. The importance of technology. Nature 209, 5019. January 8.

---. 1978a. The artifactual environment. Lenihan and Fletcher.

---. 1978b. Understanding human nature: A popular guide to the effects of technology on man and his behavior. New York: Horizon Press.

---. 1982. Technology and reality. Boston: Martinus Nijhoff.

Ferguson, Eugene S. 1977. The mind's eye: Nonverbal thought in technology. Science 179, no. 4306. August 26.

Fingar, Thomas. 1984. Consequences of catching up. In Science and technology in China special supplement. Bulletin of the Atomic Scientists 40, no. 8. October.

Fingar, Thomas, and Denis F. Simon. 1984. An overview. In Science and technology in China special supplement. Bulletin of the Atomic Scientists 40, no. 8. October.

Flavell, Richard, and Raymond Matthias. 1984. Prospects for
    transforming monocot crop plants. Nature 307, no. 5947.
    January 12.
Florman, Samuel. 1976. The existential pleasure of engineering.
    New York: St. Martin's Press.
———. 1981. Blaming technology: The irrational search for
    scapegoats. New York: St. Martin's Press.
Food and Agriculture Organization of the United Nations. 1982.
    The state of food and agriculture 1981. Rome: FAO
    Agricultural Series 142.
Foreign Affairs and National Defense Division, Congressional
    Research Service, The Library of Congress. 1984. Feeding
    the world's population: Developments in the decade
    following the world food conference of 1974. Washington,
    DC: Committee on Foreign Affairs, U.S. House of
    Representatives. October.
Fowler, William A. 1984. The quest for the origin of the
    elements. Science 226, no. 4677. November 23.
Frankel, Glen. 1984. An African success story: Despite
    drought, Zimbabwe's farmers reap crop 'miracle.' Washington
    Post. November 20.
Freeman, Orville, and Ruth Karen. 1982. The farmer and the
    money economy: The role of the private sector in the
    agricultural development of LDC's. Technological
    Forecasting and Social Change: An International Journal.
    Special Issue on The Woodlands Conference on Sustainable
    Societies: Future Roles for the Private Sector. Vol. 22,
    no. 2. October.
Frieden, Bernard J. 1979. The environmental protection hustle.
    Cambridge, MA: The MIT Press.
Friedman, Milton. 1962. Capitalism and freedom. Chicago:
    Univ. of Chicago Press.
Furst, Peter T. 1978. Spirulina. Human Nature 1, no. 3.
    March.
Cale, George. 1981. The anthropic principle. Scientific
    American 245, no. 6. December.
Galeshoff, Stuart. 1980. Triumph and failure: The American
    response to the urban water supply problem, 1850–1930. In
    Melosi, ed.
Georgescu-Roegen, Nicholas. 1971. The entropy law and the
    economic process. Cambridge, MA: Harvard Univ. Press.
———. 1975. Energy and economic myths. Southern Economic
    Journal 41. no. 3. January.
Gershaff, Stanley N. 1980. The fortification of foods. In
    Pearson and Greenwell.
Gest, Howard, and J. William Schopf. 1983. Biochemical
    evolution and energy conversion: The transition from
    fermentation to an oxygenic photosynthesis. In Schopf.
Ghosh, Pradip K., ed. 1984. Population, environment and
    resources, and third world development. Westport, CT:
    Greenwood Pass. International Development Resource Books.
    No. 5.

Gibbons, G. W., S. W. Hawking, and S. T. C. Sikos, eds. 1983.
The very early universe. Cambridge: Cambridge Univ. Press.

Giedion, Siegfried. 1948. Mechanization takes command: A
contribution to anonymous history. New York: W. W. Norton.

Giere, John P., Keith M. Johnson, and John H. Perkins. 1980. A
closer look at no-till farming. Environment 22, no. 6.
July/August.

Gifford, Roger M., J. H. Thorne, W. D. Hitz, and Robert T.
Giaquinto. 1984. Crop productivity and photoassimilate
partitioning. Science 225, no. 4664. August 24.

Glass, Billy P. 1982. Introduction to planetary geology.
Cambridge: Cambridge Univ. Press.

Goldschmidt, Walter. 1967. Man's way: A preface to the
understanding of human society. New York: Holt, Rinehart,
& Winston.

Goodall, Jane. 1963. My life among the wild chimpanzees.
National Geographic 124, no. 2. August.

Goodall, Jane, and Hugo Van Lawick. 1965. New discoveries among
Africa's chimpanzees. National Geographic 128, no. 6.
December.

Goodell, Grace E. 1984. Bugs, bonds, banks and bottlenecks:
Organizational contradictions in the new rice technology.
Economic Development and Cultural Change 33, no. 1.
October.

Goodman, G. T., and W. O. Rowe, eds. 1979. Energy risk
management. New York: Academic Press.

Grant, James P. 1984. The state of the world's children, 1984.
Oxford: Oxford Univ. Press.

Green, Harvey. 1983. The light of the home: An intimate view
of the lives of women in Victorian America. New York:
Pantheon Books.

Gregg, Alan. 1955. A medical aspect of the population problem.
Science 121, no. 3150. May 13.

Gribbin, John. 1982. Future weather and the greenhouse effect.
New York: Delacorte.

Grier, Peter. 1984. After working to limit air pollution
outdoors, U.S. looks indoors. Christian Science Monitor.
September 25.

Griffin, James, and Henry Steele. 1980. Energy economics and
policy. New York: Academic Press.

Griffin, Keith. 1972. The green revolution: An economic
analysis. Geneva: United Nations Institute for Social
Development.

Guild, Phillip W. 1976. Discovery of natural resources.
Science 91, no. 4228. February 20. Reprinted in Abelson
and Hammond.

Hahn, Caldwell. 1984. Environment and development: Balancing
both worlds. Horizons 3, no. 4. Fall.

Haldane, J. B. S. 1928. On being the right size. In Haldane,
ed., Possible worlds. New York: Harper.

Hall, A. Rupert. 1962. The scientific revolution 1500-1800:
The formation of the modern scientific attitude. 2nd

edition. London: Longmans.

Hall, David. 1984. Signposts to a sustainable world. New Scientist 102, no. 1409. May 10.

Hall, Ross Hume. 1974. Food for nought: The decline in nutrition. Baltimore: Harper & Row.

Hanson, Haldore, Norman E. Borlaug, and R. Glen Anderson. 1982. Wheat in the third world. Boulder, CO: Westview.

Hardin, Garrett. 1974. Living on a lifeboat. Bioscience 24, no. 10. October. Reprinted in The Co-Evolution Quarterly, no. 6. Summer 1975.

---. 1977. The limits of altruism: An ecologist's view of survival. Bloomington: Indiana Univ. Press.

---. 1978. Vulnerability--the strength of science. In Hardin, Stalking the wild taboo. Los Altos, CA: William Kaufman.

Hareven, Tamara K., ed. 1978. Transitions: The family and the life course in historical perspective. New York: Academic Press.

Harlan, Jack R. 1975. Crops and man. Madison, WI: American Society of Agronomy, Crop Science Society of America.

Harper, A. E. 1984. Nutrition and health in the changing environment. In Simon and Kahn.

Harris, Robert C., Christopher Hohenemser, and Robert Kates. 1979. The burden of technological hazard. In Goodman and Rowe.

Harrison, Edward R. 1981. Cosmology: The science of the universe. Cambridge, England: Cambridge Univ. Press.

Harrison, Paul. 1980. Appropriate technology: How can it reach the villages? New Scientist 88, no. 1228. November 20.

Hart, B. H. Liddell. 1965. The Liddell Hart memoirs; vol. 1, 1898-1938. New York: G. P. Putnam's Sons.

Hartley, G. S. 1976. Agricultural chemicals and the environment. In Lenihan and Fletcher.

Haskell, Thomas L. 1985. Capitalism and the origins of the humanitarian sensibility. American Historical Review (in press).

--- ed. 1984. The authority of experts: Studies in history and theory. Bloomington: Indiana Univ. Press.

Haughton, Claire Shaver. 1978. Green immigrants: The plants that transformed America. New York: Harcourt, Brace, Jovanovich.

Hausfater, Glen, and Sarah Blaffer Hrdy, eds. 1984. Infanticide: Comparative and evolutionary perspectives. New York: Aldine Publishing Co.

Hawkes, J. G. 1983. The diversity of crop plants. Cambridge: Harvard Univ. Press. Reprinted by permission. Copyright © 1983. The Harvard Univ. Press.

Hayden, F. Gregory. 1980. An assessment dependent upon technology. Journal of Economic Issues 14, no. 1. March.

Hayes, Denis. 1977. Rays of hope: The transition to a post-petroleum world. New York: W. W. Norton.

Hecht, S. B., ed. 1982. Amazonia: Agriculture and land use research. Cali, Columbia: Centre Internacional de

Agricultura Tropical (CIAT).

Heilbroner, Robert L. 1974. An inquiry into the human prospect. New York: Norton.

Hellman, Hal. 1976. Technophobia: Getting out of the technology trap. New York: M. Evans.

Henderson, Hazel. 1981. The politics of the solar age: Alternative to economics. Garden City: Anchor Press/ Doubleday.

Henig, Robin Marantz. 1983. Where surgery meets the soul. The Washington Post Book World 13, no. 16. April 17.

Hilts, Philip. 1983. Bird population of 17 million vanishes. The Guardian, March 16. International Herald Tribune, March 16.

Himowitz, Michael. 1983. U.S. seeking to ease rules on "childproof caps." Houston Chronicle. April 21.

Hinman, C. Wiley. 1984. New crops for arid lands. Science 225, no. 4669. September 28.

Hobsbaum, Eric, and Terence Ranger, eds. 1983. The invention of twentienth century art. Wall Street Journal. October 9.

Hockett, Charles F. 1960. The origins of speech. Scientific American. September.

Hoeltexhoff, Manuela. The call of the wild: 'Primitivism' in twentienth century art. Wall Street Journal. October 9.

Hofstader, Douglas R. 1983. Mathematical themes: Virus-like sentences and self-replicating structures. Scientific American 248, no. 1. January.

Holland, Heinrich. 1984. The chemical evolution of the atmosphere and oceans. Princeton, NJ: Princeton Univ. Press.

Holmberg, Scott D., Joy G. Wells, and Mitchell L. Cohen. 1984. Animal-to-man transmission of anti-microbial resistant salmonella: Investigations of U.S. outbreaks, 1971-1983. Science 255, no. 4664. August 24.

Holmberg, Scott D., Michael T. Osterholm, Kenneth A. Senger, and Mitchell L. Cohen. 1984. Drug-resistant salmonella from animals fed antimicrobials. The New England Journal of Medicine 311, no. 10. September 6.

Hopson, Janet, and Joel Guren. 1984. Diet and cancer--round 2. American Health 3, no. 8. November.

Houston Chronicle. 1980. Greenhouse effect may not be bad, says horticulturist. January 6.

Huang, H. T., and L. G. Mayfield. 1980. Biomass production and utilization. In Staples and Kuhr.

Hulse, Joseph. 1982. Food science and nutrition: The gulf between rich and poor. Science 216, no. 4552. June 18.

Hutchins, Carleen Maley. 1981. Letter to Scientific American 245, no. 4. October.

Idso, Sherwood R. 1984. Through a glass darkly. New Scientist 102, no. 1406. April 19.

Illich, Ivan. 1973. Tools for conviviality. New York: Harper & Row.

International Food Policy Research Institute (IFPRI) Annual Report. 1982. Washington, DC.

Jackson, Jean E. 1984. The impact of the state on small scale
    societies. Studies in Comparative International Development
    19, no. 2. Summer.
Jacobs, Jane. 1984. Why TVA failed. New York Review of Books
    31, no. 8. May 10.
Jäger, Jill. 1983. Climate and energy systems: A review of
    their interaction. New York: John Wiley.
Janick, Jules, Robert W. Schery, Frank W. Woods, and Vernon W.
    Ruttan. 1969. Plant science: An introduction to world
    crops. San Francisco: W. H. Freeman.
Jequier, Nicolas, ed. 1976. Appropriate technology: Problems
    and promises. Paris: Development Centre for the
    Organization for Economic Cooperation and Development.
Jefferson, Thomas. 1964. Notes on the state of Virginia. New
    York: Harper & Row.
Jha, Prem Shakar. 1980. India: A political economy of
    stagnation. Delhi: Oxford Univ. Press.
Johananson, D. C., and M. A. Edey. 1981. Lucy: The beginning
    of humankind. New York: Simon and Schuster.
Johnson, D. Gale. 1981. Conditions for more rapid agricultural
    development. In Woods.
———. 1984. World food and agriculture. In Simon and Kahn.
Joint Economic Committee, U.S. Congress. 1978. Policy and
    performance. Washington, DC: USGPO.
Jolly, Alison, and Richard Jolly. 1979. Under the zoom lens.
    Mazingira: The world forum for environment and development,
    no. 11.
Jolly, Richard, Emanuel DeKadt, Hans Singer, and Fiona Wilson.
    1973. Third world employment: Problems and strategy.
    Baltimore: Penguin Books.
Jones, Ralph, ed. 1981. Readings from Futures: A collection of
    articles from the journal Futures, 1974-80. Guildford,
    Surrey, England: Westbury Houses.
Journal of the Society for International Development, 1982. From
    exporters to importers: A new dependence. No. 4.
Kahn, Arthur D. 1970. Every art possessed by man came from
    Prometheus: The Greek tragedies and science and technology.
    Technology and Culture 11, no. 2, April. Reprinted with
    permission. Copyright © 1970. The Univ. of Chicago Press.
KaKade, M. L., and I. E. Liener. 1973. The increased
    availability of nutrients from plant foodstuffs through
    processing. In Recheigl.
Kamm, Henry. 1984. Zimbabwe beats droughts with buoyancy and
    skill. New York Times. December 2.
Kates, Robert W. 1978. Risk assessment of environmental hazard.
    New York: John Wiley & Sons.
Kellogg, Charles E. 1973. Expanding farm production: Extending
    the area of soils used. In Recheigl.
Kellogg, W. W., and R. Schware. 1981. Climate change and
    society. Boulder, CO: Westview.
Kerr, Richard A. 1984. Doubling of atmospheric methane
    supported. Science 226, no. 4677. November 23.
Kilborn, Robert, Jr. 1982. Battling for sunlight: Solar access

protection laws considered. Christian Science Monitor, reprinted in the Houston Post. January 23.

King, Elbert. 1976. Space geology: An introduction. New York: John Wiley.

King, Seth. 1984. Worldwide food production levels show a big rise. New York Times. December 2.

Kirsch, Laurence S. 1982. Behind closed doors: Indoor air pollution and government policy. Harvard Environmental Law Review 6. Reprinted by permission. Copyright © 1982. The Harvard Environmental Law Review.

Kitahara-Frisch J. 1980. Apes and the making of stone age tools. Current Anthropology 21, no. 3. July.

Knowles, Paul F. 1984. Development of new crops: Needs, procedures, strategies, and options. Ames, IA: Council for Agricultural Science and Technology, Report no. 102. October.

Kroeber, A. L. 1917. The superorganic. American Anthropologist 19, no. 2. April-June.

Lancaster, Jane B. 1968. On the evolution of tool-using behavior. American Anthropologist 70, no. 1. February.

Langer, William L. 1972. Checks on population growth, 1750-1850. Scientific American 226, no. 2. February.

---. 1974. Infanticide: A historical survey. History of Childhood Quarterly 1, no. 3. Winter.

Larson, W. E., F. J. Pierce, and R. H. Dowdy. 1983. The threat of soil erosion to long-term crop production. Science 219, no. 4584. February 4.

Lasagna, Louis. 1984. Masked men of science. The Sciences 24, no. 1. January/February.

Lasch, Christopher. 1979. Haven in a heartless world. New York: Basic Books.

Laslett, Barbara. 1980. Family membership, past and present. In Skolnick and Skolnick.

Laslett, Peter. 1965. The world we have lost: England before the industrial age. New York: Charles Scribner.

Layton, Jr., Edwin. 1974. Technology as knowledge. Technology and Culture 15, no. 1. January.

Lee, R. B., and I. De Vore, eds. 1968. Man the hunter. Chicago: Aldine Press.

Lenihan, John, and William W. Fletcher, eds. 1976. Food, agriculture, and the environment. New York: Academic Press.

Lenihan, John, and William W. Fletcher, eds. 1978. The built environment. New York: Academic Press.

Leopold, Aldo. (1949) 1966. A Sand County almanac: With essays on conservation from Round River. Reprinted. New York: Sierra Club/Ballantine.

Leopold, A. Carl, and Robert Ardrey. 1972. Toxic substances in plants and food habits of early man. Science 176, no. 4034. May 5.

Lepkowski, Wil. 1979. The social thermodynamics of Ilya Prigogine. Chemical and Engineering News 57, no. 16. April 16.

———. 1980. Science and the humanities: Bridging the gap. Chemical and Engineering News 58, no. 48. December 1.

Leroi-Gourham, Andre. 1969. Primitive societies. In Daumas.

Levin, Harold L. 1978. The Earth throughout time. Philadelphia: W. B. Saunders.

Lewantin, R. C., Steven Rose, and Leon Kamin. 1984. Not in our genes: Biology, ideology and human nature. New York: Pantheon Books.

Lewin, Roger. 1982. How did humans evolve big brains? Science 216, no. 4548. May 21.

———. 1982. Never ending race for genetic variants. Science 218, no. 4575. November 26.

———. 1983. No dinosaurs this time. Science 221, no. 4616. January 7.

Lewis, Clifford W. 1983. Global food security—a manageable problem. Development Digest 21, no. 1. July.

Lewis, John S., and Ronald G. Prinn. 1984. Planets and their atmosphere: Origins and evolution. Orlando, FL: Academic Press, International Geophysical Series, vol. 33.

Lewis, W. Arthur. 1954. Economic development with unlimited supplies of labor. The Manchester School of Economic and Social Studies 22, no. 2. May.

Linde, A. D. 1983. The new inflationary universe scenario. In Gibbons, Hawking, and Sikos.

Linton, Ralph. 1963. The study of man. New York: Appleton-Century-Crofts.

Liss, P. S., and A. J. Crane. 1984. Man made carbon dioxide and climate change: A review of the scientific problems. Norwich, England: Geo Books.

Loomis, Robert S. 1976. Agricultural systems. Scientific American 235, no. 3. September.

Lorenz, Konrad Z. 1962. King Solomon's ring. New York: Time.

Loup, Jacques. 1982. Can the third world survive? Baltimore: Johns Hopkins.

Lovelock, J. E. 1979. Gaia: A new look at life on Earth. Oxford: Oxford Univ. Press.

Lovins, Amory B. (1976) 1977. Energy strategy: The road not taken? Foreign Affairs 55, no. 1. Rev. and reprinted as chap. 2 in Lovins, Soft energy paths: Toward a durable peace. Cambridge, MA: Ballinger.

Lowie, Robert. 1940. An introduction to cultural anthropology. New York: Rinehart.

Lynd, Robert S., and Helen Merrel Lynd. 1929. Middletown: A study in American culture. New York: Harcourt and Brace.

MacDonald, Gordon J. 1982. The long-term impacts of increasing atmosphere carbon dioxide levels. Cambridge, MA: Ballinger.

MacKenzie, Debora. 1983. Seeds of conflict over food genes. New Scientist 100, nos. 1389/1390. December 22/29.

———. 1984. Man-made disaster in the Philippines. New Scientist 103, no. 1421. September 13.

Maddox, John R. 1972. The doomsday syndrome. New York: McGraw-Hill.

Mallove, Eugene F. 1984. The cosmic riddle: How rocks and
    stars became flesh and blood. The Washington Post. October
    21.
Margulis, Lynn. 1981. The origins of plant and animal cells.
    In Skinner.
Marks, Copeland, with Mintari Soeharjo. 1981. The Indonesian
    kitchen. New York: Atheneum.
Marx, Jean L. 1979. Plants: Can they live in salt water and
    like it? Science 206, no. 4423. December 7.
Mason, Edward S. (1978) 1980. Natural resources and
    environmental restrictions to growth. Challenge. January/
    February. Reprinted in Robert C. Puth, ed., Current issues
    in the American economy. Lexington, MA: D. C. Heath.
Maugh, Thomas M., II. 1979. The threat to the ozone is real,
    increasing. Science 206, no. 4423. December 7.
Mauldin, W. Parker. 1977. World population situation: Problems
    and prospects. World Development 5, nos. 5-7.
Maurice, Charles, and Charles W. Smithson. 1984. The doomsday
    myth: 10,000 years of economic crises. Stanford, CA:
    Hoover Institution Press.
Maxwell, Neville, ed. 1979a. China's road to development.
    Oxford: Pergamon.
---. 1979b. The Tachai way, part 2: The fourth mobilization.
    In Maxwell 1979a.
May, Robert M. 1973. Stability and complexity in modern
    ecosystems. Princeton, NJ: Princeton Univ. Press.
---. 1976a. Irreproducible results. Nature 262, no. 5570.
    Aug. 19.
---. 1976b. Simple mathematical models with very complicated
    dynamics. Nature 261, no. 5560. June 10.
McGee, Harold. 1984. On food and cooking: The sciences and
    lore of the kitchen. New York: Scribner.
McGill, Douglas C. 1984. What does modern art owe to the
    primitive? New York Times, Arts & Leisure Section.
    September 23.
McMahon, Thomas A., and John Tyler Bonner. 1983. On size and
    life. New York: Scientific American Library.
McNamara, Robert S. 1984. Time bomb or myth: The population
    problem. Foreign Affairs 62, no. 2. Summer.
Meadows, Donella H., Dennis L. Meadows, Jorgen Randers, and
    William W. Behrens III. (1972) 1974. The limits to growth:
    A report for the Club of Rome's project on the predicament
    of mankind. New York: Universe Books, A Potomac Associates
    Book.
Mellor, John W., and Bruce F. Johnston. 1984. The world food
    equation: Interrelations among development, employment, and
    food consumption. Journal of Economic Literature 22, no. 2.
    June.
Melosi, Martin V. 1980. Environmental crisis in the city: The
    relationship between industrialization and urban pollution.
    In Melosi, ed.
--- ed. 1980. Pollution and reform in American cities,
    1870-1930. Austin, TX: Univ. of Texas Press.

Merton, Robert K. 1949. Social theory and social structure:
    Toward the codification of theory and research. Glencoe,
    IL: The Free Press.
Miller, Barbara. 1981. The endangered sex: Neglect of female
    children in rural North India. Ithaca, NY: Cornell Univ.
    Press.
Miller, D. S. 1979. What food crises? Review of Nutrition and
    the world food crises, by Mary Alice Caliendo. Nature 281,
    no. 5728. September 27.
Miller, David H. 1981. Energy at the surface of the Earth: An
    introduction to the energetics of ecosystems. New York:
    Academic Press.
Millstone, Erik. 1984. Food additives: A technology out of
    control? New Scientist 104, no. 1426. October 18.
Minon, John H., and William H. Lawrence, eds. 1981. Legal
    aspects of solar energy. Lexington, MA: Lexington Books,
    D. C. Heath.
Moffat, Anne Simon. 1979. From the subtropical neem tree, a
    natural insecticide. The New York Times. October 9.
Mooney, Pat Roy. 1983. The law of the seed: Another
    development and plant genetic resources. Special issue of
    Development Dialogue 1-2.
Moran, Emilio, ed. 1983. The dilemma of Amazonian development.
    Boulder, CO: Westview Press.
Morawetz, David. 1977. Twenty-five years of economic
    development 1950-1975. Washington, DC: The World Bank.
    Reprinted in Finance and Development 14, no. 3. September.
    Reprinted by permission. Copyright © 1977. The World Bank.
Morrison, Philip, Phyllis Morrison, and the Office of Charles and
    Ray Eames. 1982. The powers of ten. New York: Scientific
    American Library.
Mower, Joan. 1984. Acts of God trigger disasters, but man makes
    them worse, book says. Houston Post (Associated Press news
    story). November 14.
Mumford, Lewis 1934. Technics and civilization. New York:
    Harcourt, Brace Research Council.
Murray, Patti A., and Stephen H. Zinder. 1984. Nitrogen
    fixation by a methanogenic archaebacterium. Nature 312, no.
    5991. November 15.
My. 1975. What do we use for lifeboats when the ship goes down.
    The Co-Evolution Quarterly, no. 6. Summer.
Myers, Norman. 1979. The sinking ark: A new look at the
    problem of disappearing species. Oxford: Pergamon Press.
---. 1981. Corn acquires genetic vigor from a wild relative.
    New Scientist 89, no. 1235. January 8.
---. 1984. The primary source: Tropical forests and our
    future. New York: W. W. Norton.
Nair, Kasum. 1979. In defense of the irrational peasant:
    Indian agriculture after the green revolution. Chicago:
    The Univ. of Chicago Press.
National Research Council. 1975a. The winged bean: A high-
    protein crop for the tropics. Washington, DC: National
    Academy of Science.

———. 1975b. Underexploited tropical plants with promising economic value. Washington, DC: National Academy of Science.

———. 1977a. Guayule: An alternate source of natural rubber. Washington, DC: National Academy of Science.

———. 1977b. Leucaena: Promising forage and tree crops for the tropics. Washington, DC: National Academy of Science.

———. 1977c. World food and nutrition study: The potential contributions of research. NRC study on world food and nutrition of the Commission on International Relations. Washington, DC: National Academy of Science.

———. 1979. Tropical legumes: Resources for the future. Washington, DC: National Academy of Science.

———. 1983. Changing climate: Report of the carbon dioxide assessment committee. Board on Atmospheric Sciences and Climate, Commission on Physical Science, Mathematics, and Resources. Washington, DC: National Academy Press.

———. 1984a. Amaranth: Modern prospects for an ancient crop. Washington, DC: National Academy Press.

———. 1984b. Jojoba: New crop for arid lands. Washington, DC: National Academy Press.

Needham, Joseph. Science and civilization in China. Cambridge, England: Cambridge Univ. Press (numerous volumes, still being issued).

Nevitt, Barrington. 1980. Pipeline or grapevine: The changing communication environment. Technology and Culture 21, no. 2. April.

Newall, John. 1983. Treatment for starvation may kill. New Scientist 99, no. 1374. August 18.

Newmark, Peter. 1980. Fungal food. Nature 287, no. 5777. September 4.

Nicholaides, J. J., P. A. Sanchez, D. E. Bandy, J. H. Villachia, A. J. Coutu, and C. S. Valverde. 1983. Crop production systems in the Amazon Basin. In Moran.

Nicol, Judy. 1983. 'Lost' exotic crops pushed as foods of the future. Washington Post. May 30.

Nishida, Toshisada, and Shigeo Uehara. 1980. Chimpanzees, tools, and termites: Another example from Tanzania. Current Anthropology 21, no. 5. October.

Nordhaus, William. 1974. Resources as a constraint on growth. American Economic Review 64, no. 2. May.

Noyes, Robert W. 1982. The sun, our star. Cambridge, MA: Harvard Univ. Press.

Oakley, Kenneth P. 1954. Skill as a human possession. In Singer, Holmyard, and Hall.

O'Brien, Patrick M. 1982. The function of the world market and the impact on the world food situation. Paper presented at Woodlands Conference, Houston. October 9.

Office of Technology Assessment. 1984. Technologies to sustain tropical rain forest resources. Washington, DC: Office of Technology Assessment, U.S. Government.

Oka, Takashi. 1981. "China's Sorrow" turning to joy: The

raging Yellow River is nearly tamed. Christian Science
Monitor. September 10.

O'Keefe, Phil. 1983. Combustion of a retarded technology. New
Scientist 97, no. 1350. March 24.

O'Neil, Gerard. 1974. The colonization of space. Physics Today
27, no. 9. September.

———. 1977. The high frontier: Human colonies in space. New
York: William Morrow.

O'Neil, Thomas. 1984. The primitive urge. Horizon: The
magazine of the arts 27, no. 7. September.

Opper, Jacob. 1973. Science and the arts: A study in
relationship from 1600-1900. Rutherford, NJ: Fairleigh
Dickinson Univ. Press.

Orleans, Leo A. 1984. Education, careers and social status. In
Science and technology in China special supplement.
Bulletin of the Atomic Scientists 40, no. 8. October.

Oswalt, Wendell H. 1973. Habitat and technology: The
evolution of hunting. New York: Holt, Rinehart, &
Winston.

Ourisson, Guy, Pierre Albrecht, and Michel Rohner. 1984. The
microbial origin of fossil fuels. Scientific American 251,
no. 2. August.

Paddock, William, and Paul Paddock. 1967. Famine 1975!
America's decision: Who will survive. Boston: Little,
Brown.

———. 1976. Time of famines: America and the world food crises.
(New edition of Famine 1975 with new introduction and
postscript.) Boston: Little, Brown.

Paley Commission. 1952. Resources for Freedom. Known as The
Paley Commission Report. Washington, DC: USGPO.

Palmer, Robert. 1979. A dialogue between the music of two
Indias. New York Times. September 14.

Parks, Michael. 1982. "China's Sorrow": Chinese officials
combine efforts to harness unpredictable Yellow River.
Houston Chronicle. July 3.

Pearce, Fred. 1984a. In defense of population growth. New
Scientist 103, no. 1416. August 9.

———. 1984b. Coming to terms with a carcinogen. New Scientist
104, no. 1426. October 18.

Pearce, Fred, and Jeremy Cherfas. 1984. Antibiotics breed
lethal food poisons. New Scientist 103, no. 1421.
September 13.

Pearson, Paul B., and Richard G. Greenwell, eds. 1980.
Nutrition, food, and man. Tucson, AZ: The Univ. of Arizona
Press.

Perkins, Dwight. 1977. Rural small-scale industry in the
People's Republic of China. Berkeley: Univ. of California
Press.

Perry, John S. 1984. Much ado about CO_2. Nature 31, no. 5984.
October 18.

Peters, Robert Henry. 1983. The ecological implications of body
size. New York: Cambridge Univ. Press.

Phillips, Ronald E., Robert L. Blevins, Grant W. Thomas, Wilber
    W. Frye, and Shirley H. Phillips. 1980. No-tillage
    agriculture. Science 208, no. 4448. June 6.
Pimentel, David, and Marcia Pimentel. 1979. Food, energy, and
    society. New York: John Wiley & Sons.
Pirie, N. W. 1973. Production and use of unconventional sources
    of food. In Recheigl.
---. 1976a. Food resources: Conventional and novel. London:
    Penguin Books.
---. 1976b. Using plants optimally. In Lenihan and Fletcher.
---. (1976) 1981. The world food supply: Physical limitations.
    In Jones.
Pi-Sunyer, Oriol, and Thomas R. De Gregori. 1964. Cultural
    resistance to technological change. Technology and Culture
    5, no. 2. Spring.
Ponnamperuma, Cyril. 1981. The quickening of life. In Ripley.
Popper, Karl R. 1957. The logic of scientific discovery. New
    York: Basic Books.
Population Reports. 1979a. The world fertility survey: Current
    status and findings. Series M, no. 3. July.
---. 1979b. Age at marriage and fertility. Series M, no. 4.
    November.
Postgate, J. R. 1982. The fundamentals of nitrogen fixation.
    Cambridge: Cambridge Univ. Press.
---. 1984. New kingdom for nitrogen fixation. Nature 312, no.
    5991. November 15.
Potts, Richard. 1984. Home bases and early hominids. American
    Scientist 72, no. 4. July–August.
Poundstone, William. 1984. The recursive universe: Cosmic
    complexity and the limits of scientific knowledge. New
    York: William Morrow.
Prather, Michael J., Michael B. McElroy, and Steven C. Wofsky.
    1984. Reductions in ozone at high concentrations of
    stratospheric halogens. Nature 312, no. 5991. November 15.
Priestly, J. B. 1960. Literature and Western man. New York:
    Harper & Brothers.
Prigogine, Ilya. 1980. From being to becoming: Time and
    complexity in the physical sciences. San Francisco: W. H.
    Freeman.
Prigogine, Ilya, and Isabelle Stengers. 1984. Order out of
    chaos: Man's new dialogue with nature. New York: Bantam
    Books.
Public Health Service. 1979. Health United States. Washington,
    DC: Office of Health Research, Statistics, and Technology,
    U.S. Dept. of Health, Education, and Welfare.
Rand McNally. 1979. Our magnificent Earth: A Rand McNally
    atlas of Earth resources. Chicago: Rand McNally.
Randhawa, Mohindar Singh. 1974. Green revolution. New York:
    John Wiley & Sons.
Rao, C. H. Hanamantha. 1975. Technological change and
    distribution of gains in India. Delhi: MacMillan, for the
    Institute of Economic Growth.

Rao, Radhakrishna. 1980. When alternatives are inappropriate. New Scientist 86, no. 1201. April 3.

Raven, Peter H. 1984. Third world in the global future. Bulletin of the Atomic Scientists 40, no. 9. November. (Adapted from a symposium, Knockdown-Dragout in the Global Future, American Association for the Advancement of Science, New York).

Rawski, Thomas G. 1979. Economic growth and employment in China. New York: Oxford Univ. Press.

---. 1980. China's transition to industrialism: Producer goods and economic development in the twentieth century. Ann Arbor: The Univ. of Michigan Press.

Ream, Lloyd W., and Milton P. Gordon. 1982. Crown gall disease and prospects for genetic manipulation of plants. Science 218, no. 4575. November 26.

Recheigl, Miloslav, Jr., ed. 1973. Man, food and nutrition: Strategies and technological measures for alleviating the world food problem. Cleveland: CRC Press.

Reeves, Hubert. 1984. Atoms of silence: An exploration of cosmic evolution. Trans. by Ruth A. Lewis and John S. Lewis. Cambridge, MA: The MIT Press.

Rensberger, Boyce. 1984. Two 'natural disasters' laid to humans: Misuse of land blamed in Ethiopia and Bangladesh. Washington Post. November 14.

Repetto, Robert. 1979. Economic equality and fertility in developing countries. Baltimore: The Johns Hopkins Univ. Press for Resources for the Future.

Revelle, Roger. 1963. Water. Scientific American 209, no. 3. September.

---. 1976. The resources available for agriculture. Scientific American 235, no. 3. September.

Rhodes, Martha E. 1979. The 'natural' food myth. The Sciences 19, no. 5. May/June.

Richard, Paul. 1984. Magical affinities: Linking tribal and Western art at the Museum of Modern Art. Washington Post. September 30.

Riding, Alan. 1984. For Ecuador Indians, pride and profit in weaving. New York Times. May 14. Reprinted by permission. Copyright © 1984. The New York Times Company.

Rifkin, Jeremy. 1980. Entropy: A new world view. New York: Viking.

Ripley, Dillon S., ed. 1981. Fire of life. New York: Smithsonian Exposition Books.

Riskin, Carl. 1979. Intermediate technology in China's rural industries. In Robinson.

Robbins, Don. 1984. Preparing the past for the future. New Scientist 104, no. 1426. October 18.

Robertson, Miranda. 1983. The ti plasmid as genetic engineer. New Scientist 98, no. 1359. May 26.

Robinson, Austin, ed. 1979. Appropriate technology for third world development. New York: St. Martin's Press.

Robinson, Mary. 1981. From Southwestern deserts, jojoba plant--

a potential cornucopia. The Christian Science Monitor. October 16.

Robinson, R., and P. Johnstone, eds. 1972. Prospects for employment opportunities in the 1970's. London: H.M.S.O.

Rockwood, Walt. 1983. "New" biotechnology in international agricultural development. Horizons (Agency for International Development) 2, no. 10. November.

Rosa, Nicholas. 1982. The origins of life. Oceans 15, no. 5.

Rosenberg, Nathan. 1973. Innovative responses to materials shortages. American Economic Review 63, no. 2. May.

———. 1980. Historical relations between energy and economic growth. In Dunkerly.

———. 1982. Inside the black box: Technology and economics. Cambridge, England: Cambridge Univ. Press.

Rubin, William. 1984. "Primitivism" in twentieth century art. New York: Musuem of Modern Art. Vol. 102.

Russell, Cristine. 1984. USDA using human gene in effort to grow super livestock. Washington Post. October 1.

Russell, John. 1984. Primitive spirits invade the modern. New York Times. September 28.

Ruttan, Vernon W. 1982. Agricultural research policy. Minneapolis: Univ. of Minnesota Press.

———. 1983. The global agricultural support system. Science 222, no. 4619. October 7.

Ruttan, Vernon W., and Hans P. Binswanger. 1978. Induced innovation and the green revolution. In Binswanger and Ruttan.

Ruttan, Vernon W., and Yujiro Hayami. 1973. Technology transfer and agricultural development. New York: Agricultural Development Council.

Rybcznski, Witold. 1983. Taming the tiger: The struggle to control technology. New York: Viking Press.

Sahlins, Marshall D., and Elman R. Service, eds. 1980. Evolution and culture. Ann Arbor, MI: Univ. of Michigan Press.

Salam, Abdus. 1983. Technology for development: How the third world can benefit from Western exports. Asian Post 2, no. 4. November 5.

Salas, Raphael. 1984. Reflections on population. Elmsford, N.Y.: Pergamon Press.

Salati, Eneas, and Peter B. Vose. 1984. Amazon basin: A system in equilibrium. Science 225, no. 4658. July 13.

Sale, Kirkpatrick. 1980. Human scale. New York: Coward, McCann & Geoghegan.

Sanchez, Pedro, Dale E. Bandy, J. Hugo Villachica, and John J. Nicholaides. 1982. Amazon basin soils: Management for continuous crop production. Science 216, no. 4548. May 21.

Sanchez-Albornez, Nicholas. 1974. The population of Latin America. Berkeley: Univ. of California Press.

Sanders, William T., Jeffrey R. Parsons, and Robert S. Santley. 1979. The basin of Mexico: Ecological process in the evolution of a civilization. New York: Academic Press.

Saxena, R. C. 1983. Naturally occurring pesticides and their
potential. In Shemilt.

Schelling, Thomas G. 1983. Climatic change: Implications for
welfare and policy. In National Research Council.

———. 1984. Anticipating climate change. Environment 26, no. 8.
October.

Schiff, H. I. 1984. Ozone fears revisited. Nature 312, no.
5991. November 15.

Schmidt-Nielson, Knut. 1984. Scaling: Why is animal size so
important? Cambridge, England: Cambridge Univ. Press.

Schneider, Piere. 1980. Is science always at service of art?
New York Times. December 30.

Schneider, Stephen H. 1976. The genesis strategy: Climate and
global survival. New York: Plenum Press.

Schneider, Stephen H., and Rondi Londer. 1984. The coevolution
of climate and life. San Francisco: Sierra Club Books.

Schonberg, Harold C. 1978. Review of The orchestra: A history,
by Henry Raynor. The New York Times Book Review. June 18.
Reprinted by permission. Copyright © 1978. The New York
Times Company.

Schopf, J. William, ed. 1983. Earth's earliest biosphere: Its
origin and evolution. Princeton, NJ: Princeton Univ.
Press.

Schultz, Theodore W. 1965. Transforming traditional
agriculture. New Haven, CT: Yale Univ. Press.

———. 1981. Investing in people: The economics of population
quality. Berkeley & Los Angeles: Univ. of California
Press.

———. 1984. The dynamics of soil erosion in the United States.
In Baden.

Schultz, Theodore, and others. 1977. Lectures in agricultural
economics. Washington, DC: Economic Research Service.

Schumacher, E. F. 1973. Small is beautiful: Economics as if
people mattered. New York: Harper & Row.

Scobie, Grant M. 1979. Investment in international agricultural
research: Economic dimensions. World Bank Staff Working
Paper no. 361. Washington, DC: World Bank.

Scobie, James. 1964. Revolution on the Pampas: A social
history of Argentina wheat, 1860-1910. Austin, TX: The
Univ. of Texas Press for the Institute of Latin American
Studies.

Scott, Anne. 1983. Boosting corn with a tough cousin from the
hills. South, no. 33. July.

Scott, Tom K., ed. 1979. Plant regulation and world
agriculture. New York: Plenum.

Scrimshaw, Nevin S. 1984. The politics of starvation.
Technology Review 87, no. 6. August/September.

Scrimshaw, Nevin S., and Lance Taylor. 1980. Food. Scientific
American 243, no. 3. September.

Sebestik, Jan. 1983. The rise of technological science.
History and Technology 1, no. 1.

Seckler, David. 1980. Malnutrition: An intellectual odyssey.

Western Journal of Agricultural Economics 5, no. 2. December.

Sedjo, Roger A., and Marion Clawson. 1984. Global forests. In Simon and Kahn.

Seeds, Michael A. 1981. Horizons. Belmont, CA: Wadsworth.

Seielstad, George. 1983. Cosmic ecology: The view from the outside in. Berkeley and Los Angeles: Univ. of California Press.

Sen, Amartya. 1981. Poverty and famines: An essay on entitlement and deprivation. Oxford: Clarendon.

Seneviraine, Gammi. 1983. New facts of life: Spreading the Biotech revolution. South, no. 32.

Shabecoff, Philip. 1984. Natural disasters: Study says man, not nature, is to blame. New York Times. November 18.

Sheldon, Richard C. 1982. Phosphate rock. Scientific American 246, no. 4. June.

Shemilt, L. W., ed. 1983. Chemistry and world food supplies: The new frontiers. Oxford and New York: Pergamon.

Shepard, James F. 1982. The regeneration of potato plants from leaf-cell protoplasts. Scientific American 246, no. 5. May.

Shepard, James F., Dennis Bidney, and Elias Shahin. 1980. Potato protoplasts in crop improvement. Science 208, no. 4439. April 4.

Shorter, Edward. 1971. Infanticide in the past. Review of Slaughter of the innocents, by Bakan. History of Childhood Quarterly 1, no. 1. Summer.

———. 1982. A history of women's bodies. New York: Basic Books.

Shu, Frank H. 1982. The physical universe: An introduction to astronomy. Mills Valley, CA: University Science Books.

Sigurdson, Jon. 1977. Rural industrialization in China. Cambridge, MA: Harvard Univ. Press.

———. 1980. Technology and science in the People's Republic of China. Oxford: Pergamon.

Simberloff, Daniel. 1984. The great god of competition. The Sciences 24, no. 4. July–August.

Simmons, Harvey. 1974. System dynamics and technocracy. In Cole et al.

Simon, Denis F. 1984. International influences. In Science and technology in China special supplement. Bulletin of the Atomic Scientists 40, no. 8. October.

Simon, Julian L. 1980. Resources, population, environment: An oversupply of false bad news. Science 208. June 27.

———. 1981. The ultimate resource. Princeton, NJ: Princeton Univ. Press.

———. 1984a. Myths of overpopulation. Wall Street Journal. August 3.

———. 1984b. Bright global future. Bulletin of the Atomic Scientists 40, no. 9. November. (Adopted from a symposium, Knockdown-Dragout on the Global Future, American Association for the Advancement of Science, New York.)

Simon, Julian L., and Kahn, Herman, eds. 1984. The resourceful Earth: A response to global 2000. New York: Basil Blackwell.

Simon, Julian L., and Wildavsky, Aaron. 1984. On species loss, the absence of data and risks to humanity. In Simon and Kahn.

Sinclair, T. C. 1974. Environmentalism: A la recherche du temps perdu--biens perdu? In Cole et al.

Singer, Charles, E. J. Holmyard, and A. C. Hall, eds. 1954. A history of technology. London: Oxford Univ. Press.

Sinha, Radha, ed. 1978. The world food problem: Consensus and conflict. Oxford: Pergamon.

Skinner, Brian J., ed. 1981. Paleontology and paleoenvironments. Los Altos, CA: William Kaufman.

Skolnick, Arlene. 1984. Today's family: Myths and realities. Journal of Reformed Judaism 31, no. 1. Winter.

Skolnick, Arlene, and Jerome H. Skolnick (eds.). 1980. The family in transition: Rethinking marriage, child rearing and family organization. (3rd edition). Boston: Little, Brown and Co.

Slater, Lloyd E. 1981. Three approaches to reducing climate's impact on food supplies. In Slater and Levin.

Slater, Lloyd E., and Susan K. Levin, eds. 1981. Climate's impact on food supplies: Strategies and technologies for climate defensive food production. Boulder, CO: Westview.

Slessor, Malcolm. 1976. Energy requirements in agriculture. In Lenihan and Fletcher.

Slessor, Malcolm, and Chris Lewis. 1979. Biological energy sources. London: E&FN Span Ltd.

Smalley, E. V. 1893. The isolation of life on prairie farms. Atlantic Monthly 72, no. 431. September.

Smil, Vaclav. 1979a. Intermediate energy technology in China. In Maxwell.

---. 1979b. Renewable energies: How much and how renewable? The Bulletin of the Atomic Scientists 35, no. 1. December.

---. 1983. Biomass energies: Resources, links, constraints. New York: Plenum Press.

---. 1984. The bad Earth: Environmental degradation in China. New York: M. E. Sharpe, Inc.

Smith, Cyril Stanley. 1982. A search for structure. Cambridge, MA: MIT Press.

Smith, David G., ed. 1981. The Cambridge encyclopedia of Earth sciences. New York: Crown Inc./Cambridge Univ. Press.

Smith, Irene M. 1982. Carbon dioxide--emissions and effects. London: IEA Coal Research, Report no. ICTIS/TR18.

Smith, Nigel. 1983a. Triticale: The birth of a new cereal. New Scientist 97, no. 1340. January 13.

---. 1983b. New genes from wild potatoes. New Scientist 98, no. 1359. May 26.

Smith, V. Kerry, ed. 1978. Scarcity and growth reconsidered. Baltimore: Johns Hopkins.

Smith, Vallence L. 1977. Hosts and guests: The anthropology of

tourism. Philadelphia: Univ. of Pennsylvania Press.
Solow, Robert. 1976. Lecture given at Univ. of Houston to
    Economics faculty. April.
Spengler, John P., and Ken Sexton. 1983. Indoor air pollution:
    A public health perspective. Science 221, no. 4605. July
    1.
Spier, F. G. 1970. From the hand of man: Primitive and
    preindustrial technologies. Boston: Houghton, Mifflin.
Sprague, G. F. 1973. Increasing crop yields: Technical
    measures for increasing productivity. In Recheigl.
Stanley, Steven M. 1981. The new evolutionary timetable:
    Fossils, genes and the origin of species. New York: Basic
    Books.
Staples, Richard C., and Ronald J. Kuhr, eds. 1980. Linking
    research to crop production. New York: Plenum.
Stavis, Benedict. 1978. The politics of agricultural
    mechanization in China. Ithaca: Cornell Univ. Press.
Stern, Curt. 1955. Qualitative aspects of the population
    problem. Science 121, no. 3150. May 13.
Stewart, Frances. 1974. Technology and employment in LDC's. In
    Edwards.
---. 1977. Technology and underdevelopment. Boulder, CO:
    Westview Press.
Stewart, Frances, and Paul R. Streeten. 1973. Conflicts between
    output and employment objectives. In Jolly et al.
Stini, William A. 1980. Human adaptability to stress. In
    Pearson and Greenwell.
Stobaugh, Robert, and Daniel Yergin, eds. 1979. Energy future:
    The report of the Harvard Business School energy project.
    New York: Random House.
Stone, Lawrence. 1977. The family, sex, and marriage in England
    1500-1800. New York: Harper & Row.
Strasser, Susan. 1982. Never done: A history of American
    housework. New York: Pantheon.
Strassmann, W. Paul. 1965. Technological change and economic
    development: The manufacturing experience of Mexico and
    Puerto Rico. New York: Cornell Univ. Press.
Street, James H., and Dilmus D. James, eds. 1979. Technological
    progress in Latin America: Prospects for overcoming
    dependency. Boulder, CO: Westview.
Sun, Marjorie. 1984a. In search of salmonella's smoking gun.
    Science 226, no. 4670. October 5.
---. 1984b. Use of antibiotics in animal feed challenged.
    Science 226, no. 4671. October 12.
Suttmeier, Richard P. 1984. New conflicts in the research
    environment. In science and technology in China special
    supplement. Bulletin of the Atomic Scientists 40, no. 8.
    October.
Swaminathan, M. S. 1983. Agricultural progress: Key to third
    world prosperity. Third World Quarterly 5, no. 3. July.
---. 1984. Rice. Scientific American 250, no. 1. January.
Tannahill, Reay. 1973. Food in history. New York: Stein and
    Day.

Tarr, Joel A. 1971. Urban pollution—many long years ago. American Heritage 22, no. 6. October.

Tekinel, Osman. 1979. Water stress and its implications (irrigation) in the future of agriculture. In Scott.

Teply, L. J. 1973. Food fortification. In Recheigl.

Thomas, Keith. 1983. Man and the natural world: A history of the modern sensibility. New York: Pantheon.

Timberlake, Lloyd, Jon Tinker, Barbara Cheney, and John McCormick. 1982. Saving the world's genetic bank: The benefit of the Earth's "wild" resources. World Press Review 29, no. 12. December.

Timmer, C. Peter, Walter P. Falcon, and Scott R. Pearson. Food policy analysis. Baltimore: The Johns Hopkins Univ. Press.

Timmer, C. Peter, J. W. Thomas, J. T. Wells, and D. Morawetz. 1975. The choice of technology in developing countries: Some cautionary tales. Cambridge, MA: Harvard Univ. Press (Center for International Affairs).

Tonge, Peter. 1979. Tasty amaranth grain has solid potential. The Christian Science Monitor. September 21.

———. 1982. Incredible edible bean? The Christian Science Monitor. April 30.

Trefil, James S. 1983. The moment of creation: Big bang physics from before the first millisecond to the present universe. New York: MacMillan.

Tucker, Jonathan R. 1983. Appropriate technology: The Lorena solution of the firewood crisis. Environment 25, no. 3. April.

Tudge, Colin. 1980. Brains for energy. New Scientist 85, no. 1192. Jan. 31.

———. 1983. The future of crops. New Scientist 98, no. 1359. May 26.

Turner, Frederick. 1984. Escape from modernism: Technology and the future of the imagination. Harper's 269, no. 614. November.

Uhlenberg, Peter. 1978. Changing configurations of the life course. In Hareven.

Usher, Abbott Payson. 1959. A history of mechanical inventions. Boston: Beacon.

Vajk, Peter J. 1978. Doomsday has been canceled. Culver City, CA: Peace Press.

Van Heyningen, W. E., and John R. Seal. 1980. Cholera: The American scientific experience, 1947–1980. Boulder, CO: Westview Press.

Veblen, Thorstein. (1906) 1961. The place of science in modern civilization. Journal of Sociology 11, no. 5. Reprinted in Veblen, The place of science in modern civilization and other essays. New York: Russell & Russell.

———. 1922. The instinct of workmanship. New York. B. W. Heubsch.

Vietmeyer, Noel D. 1981. Rediscoving America's forgotten crops. National Geographic 15, no. 5. May.

Villareal, Reuben L. 1980. Linking basic research to

improvement programs in developing countries. In Staples
and Kuhr.
Von Neumann, John. 1966. Theory of self-reproducing automata.
Urbana and Chicago: Univ. of Illinois Press.
Von Weizsäcker, E. U., M. S. Swaminathan, and Aklila Lemma, eds.
1983. New frontiers in technology applications:
Integration of emerging and traditional technologies.
Dublin, Ireland: Tycooly International Publishing Ltd.
Wade, Nicholas. 1979. CO_2 in climate: Gloomsday predictions
have no fault. Science 206, no. 4421. November 23.
Waggoner, Paul E. 1983. Agriculture and a climatic change by
more carbon dioxide. In National Research Council.
Waldrop, M. Mitchell. 1984. An inquiry into the state of the
Earth. Science 226, no. 4670. October 5.
Walsh, John. 1981. Genetic vulnerability down on the farm.
Science 214, no. 4517. October 9.
———. 1984. Seeds of dissension sprout at FAO. Science 228, no.
4682. January 13.
Washburn, Mark. 1981. In the light of the sun. New York:
Harcourt Brace Jovanovich.
Washburn, Sherwood. 1960. Tools and human evolution.
Scientific American 203, no. 3. September.
Webb, Walter Prescott. 1931. The great plains. New York: Ginn
and Company.
Weber, E. J. 1978. New beginnings for an ancient crop. IDRC
Reports 7, no. 1. March.
Webster, Bayard. 1983. Forest's role in weather documented in
the Amazon. New York Times. July 5.
Weinberg, Steven. 1977. The first three minutes. New York:
Basic Books.
Weins, Thomas R. 1978. The evolution of policy and capability
in China's agricultural technology. In Joint Economic
Committee, U.S. Congress, Chinese economy post-Mao.
Welker, Robert Henry. 1955. Birds and men: American birds in
science, art, literature and conservation, 1800–1900.
Cambridge, MA: Harvard Univ. Press.
Wenk, Edward, Jr. 1979. Margins for survival: Overcoming
political limits in steering technology. Oxford & New York:
Pergamon.
Wenke, Robert J. 1980. Patterns in prehistory: Mankind's first
three million years. New York: Oxford Univ. Press.
Whelan, Elizabeth M. 1980. Beyond the labels. New York Times.
January 14.
Whisnant, David E. 1983. All that is native and fine: The
politics of culture in an American region. Chapel Hill, NC:
Univ. of North Carolina Press.
White, Leslie O. 1959. The evolution of culture. New York:
McGraw-Hill.
White, Lynn, Jr. 1962. Medieval technology and social change.
New York: Oxford Univ. Press.
Whitney, Eleanor Noss, and Eva May Nunnelly Hamilton. 1981.
Understanding nutrition, 2nd ed.. St. Paul, MN: West
Publishing Co.

Wijkman, Anders, and Lloyd Timberlake. 1984. Natural disasters: Acts of God or acts of man? Washington, DC: An Earthscan Paperback, published by the International Institute for Environment and Development.

Wilford, John Noble. 1980. Agriculture meets the desert on its own terms. New York Times. January 15.

Wittwer, Sylvan H. (1977) 1978. Assuring our food supply: Technology resources and policy. World Development 5, nos. 5-7. Repr. in Sinha. Reprinted by permission from Sinha, The World Food Problem, copyright © 1978. Pergamon Press, Ltd.

Wolman, Yecheskel. 1981. Origin of life. Dordrecht, The Netherlands: D. Reidel Publishing.

Woodell, C. M., J. E. Jobbie, J. M. Melillo, B. Moore, B. J. Peterson, and G. R. Shaver. 1983. Global deforestation: Contribution to atmospheric carbon dioxide. Science 222, no. 4628. December 9.

Woods, Richard G., ed. 1981. Future dimensions of world food and population. Boulder, CO: Westview.

World Bank. 1979. World Development Report 1979. New York: Oxford Univ. Press.

———. 1980. Health: Sector policy paper, 2nd ed. Washington, DC: World Bank.

———. 1984. World development report 1984. New York: Oxford Univ. Press.

World Development. 1977. The choice of technology in developing countries. Special issue vol. 5, nos. 9/10. September/October.

Young, James Harvey. 1961. The toad stool millionaires: A social history of patent medicines in America before federal regulation. Princeton, NJ: Princeton Univ. Press.

Zamora, Lois P., ed. 1982. The apocalyptic vision in America. Bowling Green, Ohio: Bowling Green Univ. Popular Press.

Zimmermann, Erich W. 1951. World resources and industries: A functional appraisal of the availability of agricultural and industrial materials. New York: Harper & Brothers.

Zvelebil, Mark. 1984. Clues to recent human evolution from specialized technologies. Nature 307, no. 5949. January 26.

AUTHOR INDEX

SUBJECT INDEX

Thomas R. DeGregori is a professor of economics at the University of Houston, where he is also a research associate for the Institute for International Business Analysis. He also serves as adjunct professor at St. Thomas University in Houston and as executive associate for international programs at Denver Research Institute. DeGregori is the author of several books, monographs, and reports on economic development as well as the author of two dozen articles and numerous book reviews and conference papers. Most of his work is concerned with technology and its role in society and development. DeGregori is currently working with an engineer, Dr. Jack Matson, and with a project design and evaluation officer, Randal Thompson, to translate the principles of technology in this book into an Expert Systems Technology Transfer computer program.